* TC 274
OC

AF873891

K. Frank P. Petersen

Historic Preservation in the USA

Springer
Berlin
Heidelberg
New York
Barcelona
Hong Kong
London
Milan
Paris
Tokyo

Karolin Frank
Patricia Petersen (Editor)

Historic Preservation in the USA

with 62 Figures and 19 Tables

Springer

DR. KAROLIN FRANK
Georgiweg 37
22453 Hamburg
Germany
EPost@karolin-frank.de

Geographisches Institut
Universität Heidelberg
125/02
Bibliothek

PROFESSOR PATRICIA PETERSEN
University of Toronto
Innis College
2 Sussex Avenue
Toronto, Ontario M1N 2T1
Canada

Translated by Hannah M. Mowat, Jeff Smith, and Karolin Frank

Cover Photo by Karolin Frank: Savannah Historic District, Savannah, GA, USA, May 1999

ISBN 3-540-41735-4 Springer-Verlag Berlin Heidelberg New York

Library of Congress Cataloging-in-Publication Data applied for

Die Deutsche Bibliothek – CIP Einheitsaufnahme

Historic preservation in the USA : with 19 tables / Karolin Frank ; Patricia Petersen (ed.). Transl. by Hannah M. Mowat and Jeff Smith. - Berlin ; Heidelberg ; New York ; Barcelona ; Hong Kong ; London ; Milan ; Paris ; Tokyo : Springer, 2002
ISBN 3-540-41735-4

This work is subject to copyright. All rights are reserved, whether the whole or part of the material is concerned, specifically the rights of translation, reprinting, reuse of illustrations, recitation, broadcasting, reproduction on microfilm or in any other way, and storage in data banks. Duplication of this publication or parts thereof is permitted only under the provisions of the German Copyright Law of September 9, 1965, in its current version, and permission for use must always be obtained from Springer-Verlag. Violations are liable for prosecution under the German Copyright Law.

Springer-Verlag is a member of BertelsmannSpringer Science+Business Media GmbH

http://www.springer.de

© Springer-Verlag Berlin Heidelberg 2002
Printed in Germany

The use of general descriptive names, registered names, trademarks, etc. in this publication does not imply, even in the absence of a specific statement, that such names are exempt from the relevant protective laws and regulations and therefore free for general use.

Product liability: The publishers cannot guarantee the accuracy of any information about the application of operative techniques and medications contained in this book. In every individual case the user must check such information by consulting the relevant literature.

Camera ready : Guido Oesterlein, Tettnang
Cover design: E. Kirchner, Heidelberg
Printed on acid-free paper SPIN 10793223 30/3130/as 5 4 3 2 1 0

To my parents

To the United States

America, you're better off
Than our continent that's old.
At tumbled-down castles you scoff,
You lack basalt, I'm told.
Within, nothing daunts you
In times rife with life,
No memory haunts you
Nor vain, idle strife. […]

Den Vereinigten Staaten

Amerika, du hast es besser
Als unser Kontinent, der alte.
Hast keine verfallene Schlösser
Und keine Basalte.
Dich stört nicht im Inneren,
Zu lebendiger Zeit,
Unnützes Erinnern
Und vergeblicher Streit. [..]

Johann Wolfgang von Goethe,
Xenien (ca. 1796)

Preface by the Author

"Is there such a thing as historic preservation in the U.S.A.?" It is above all this question, reflecting the cultural skepticism towards the U.S.A., still evident in Europe today, and the obstinate belief that "historic preservation cannot exist" in the United States that reinforced my desire to examine and analyze the many existing structures and methods of urban historic preservation that can be found in the U.S.A.

I became first interested in carrying out a specialist regional study on the United States during a year spent studying at Temple University in Philadelphia, Pennsylvania from 1994-1995, funded by a DAAD (German Academic Exchange Service) scholarship. Subsequent periods spent in the United States led me to realize the extent of the cultural differences and prejudices that separate Germany and the U.S.A., most specifically within the field of historic preservation.

The issues and findings led to this book based on research in Philadelphia, Boston, charleston und Savannah. Thank to the support of the following people, I was able to publish this book.

First of all, I would like to thank my much-appreciated advisor (*Doktorvater*), Professor Dr. Frank Norbert Nagel, who oversaw the writing of this study and was a constant presence throughout my studies, always willing to answer my queries and offer academic support.

I also thank Professor Dr. Hermann Korte for supporting my study by acting as a second referee and for his invaluable advice on many aspects of sociology. I should furthermore like to express my gratitude to Professor Dr. Dr. h.c. Elisabeth Lichtenberger, who also acted as a referee for this study and allowed me to benefit from her expertise.

Particular thanks go to Professor Patricia Petersen, Ph.D. from the University of Toronto, who has kindly agreed to edit this book. I would also like to thank Professor Robin Williams, Ph.D. from the Savannah College of Art and Science who kindly wrote the foreword for this manuscipt.

During research periods spent in the United States, many people and organizations gave me the benefit of their time and expertise. These include the numerous archives, libraries, voluntary workers from the neighborhood organizations, those actively involved in programs of historic preservation in the U.S.A. and all interviewees. Thank you very much!

I am grateful to Jost „Alonzo" Mühlbach for sharing his knowledge on EDV questions and more generally for always being there when needed. Due to a tragic accident Jost is not able to see this finished book, but I can only hope he knows how much I appreciate his support on this work.

Finally, I should like to thank the following -alphabeticially listed-: Dr. Asif Ahmad, Dr. Ulf Balack (for patiently drawing up the index), Jan Behrens, Johannes Broermann, Thomas Devaney, Marion Dohr, Dr. Götz Goldammer, Professor Dr. E. Grimmel, Dr. Jakob Kim, *Doktorkollege* Dr. Joachim Weber and many others who have helped at one time or another with my work. Every single one of these people has contributed in his or her own way towards this book!

Last but not least, a special thank you to my parents, Friederike and Robert Frank, who were constantly there for me with a seemingly endless supply of patience and understanding, always ready to offer help and advice when it was most needed. This book is dedicated to them both.

Hamburg, Germany, September 2001 Karolin Frank

Foreword

Karolin Frank, *Historic Preservation in the USA*
By Robin B. Williams, Ph.D.

Originating out of a desire to save remarkable individual buildings of historical significance, historic preservation in the United States has developed into a movement that is shaping contemporary urban planning and architectural design decisions. From its isolated and sporadic origins in the hands of a relatively few, the historic preservation movement has grown into a popular force, involving a greater number of people in questions relating to the design of the built environment than ever before. The popularity of preservation has also greatly broadened American public and professional interest in the country's architectural history.

Architecture in the United States has historically been subject to greater pressure to change than is experienced in Europe, where continuity of built form has been largely impervious to the processes of modernization and the more turbulent political upheavals. At times, the American embrace of economic development was unsentimentally pragmatic. As older buildings lost their immediate use or gravitated towards obsolescence, few questioned the logic of replacing them with newer, more efficient buildings. Although many European cities suffered tremendous damage as a result of the extensive bombing during World War II, many North American cities witnessed even greater architectural destruction following the war due to the craze for new development. As a result, the number of American cities that possess a sense of historical integrity on an urban level are relatively few. It is no surprise that three of the best preserved historic cities in the country – New Orleans, Savannah, and Charleston, SC – are in the south, where industrialization was much less intense than in the north.

Yet, Americans since the country's founding have also been keenly aware of their history – at least those parts easily woven into a national mythology. One suspects the American compulsion to celebrate or commemorate certain events or figures from the past more energetically than do the citizens of other western countries (and I speak specifically from the perspective of a Canadian who has also lived in Europe) derives from a patriotic ambition to establish – as quickly as possible – a powerful national identity that effectively competed with those of European nations. An interest in history driven by patriotism, however, fostered a selective approach to preservation. Generally, sites and buildings were preserved if they could be related to some significant historical event or figure, most notably George Washington. Sites and buildings were treated like relics: they were preserved, but usually also severed from their historical surroundings. A good example is the house of Margaret Mitchell in Atlanta, where the author wrote *Gone with the Wind*. It stands alone, surrounded by vacant lots and some modern high-rises, and utterly bereft of the residential neighborhood of which it was originally a part.

Since World War II, the pace of change in America has steadily accelerated, most acutely in the nation's cities. This process helped fuel the dramatic growth of interest in the preservation movement, as citizens felt an urgency to retain links to the past. An interest in the continuity of traditions has grown to rival the long-standing embrace of progress. With the codification of this interest into law with the Historic Preservation Act of 1966, widespread support for preservation and an interest in American architectural history in general has blossomed.

The evolution of architectural history as a field offers telling evidence of the changes in post-war interests, paralleling the rise of the preservation movement. Prior to the mid-1960s, the study of architectural history took place within departments of art history and typically focused almost exclusively on European monuments. Buildings were examined in isolation like paintings, free from their broader physical context. As the interest in American architectural traditions grew, so too did the desire to study so-called vernacular architecture – the large majority of buildings in the United States erected not by trained architects but by largely anonymous and often non-professional builders and craftsmen. The process of identifying, researching, documenting and nominating buildings and districts to the National Register of Historic Places served as an important catalyst, providing an important incentive to take notice of an ever-increasing amount of the America's built heritage. The study of architectural history in academia continues to broaden its horizons, embracing the methodologies of folk historians, social and cultural anthropologists, archeologists and geographers. This interdisciplinary cross-fertilization has been greatly assisted by the growing importance of historic preservation.

Outside academia, too, architectural history has recently benefited from the popularity of preservation. A rapidly growing number of professional positions for architectural historians, especially in the areas of historic site management, preservation planning, main street management, and historic foundation management, would not have developed without the establishment of the preservation movement.[1]

The blossoming of the historic preservation movement has likewise influenced the field of contemporary architecture, although one would find little evidence of this within the studios of most schools of architecture. Rather, one must look to the growing number of residential developments and renovation projects of existing buildings to find palpable evidence of a strong interest in history. While many

[1] For example, in 2000, there were over 300 advertised full-time positions for architectural historians in the United States. Research on these numerous non-academic career opportunities for architectural historians has been compiled by my department of Architectural History at the Savannah College of Art and Design. Summary statistical results of the survey are posted at the department web site: www.scad.edu/archhist/careers.html.

contemporary architects decry the supposedly derivative and inauthentic evocations of the past as another wave of shallow historicism, countless homeowners willingly support the efforts of architects and developers of new urbanist or neo-traditionalist designs.

The gulf existing between most contemporary architects and historic preservationists also separates contemporary urban designers and city planners from the traditional tenets of city building found in most historic cities. It is a sobering thought to consider that downtown Savannah, with its many historic squares, generous streets, relatively small and densely built property lots would not conform to the zoning guidelines by which the more recent parts of that city are being built. In other words, were downtown to be designed and built today, it would not be legal. And yet, the downtown area of Savannah, particularly the area defined by the plan originated by Oglethorpe, the founder of Georgia in 1733, attracts the attention of urbanists, architects, and tourists from around the world. Few flock to the city's southside to admire the suburban sprawl or predictable malls, outlet stores and generic houses.

What is the lesson to be learned? It is obvious that Americans need to adopt a more comprehensive and integrated approach to the designing of their cities, one that both incorporates the past as a living element and that learns from the past for the best of its design guidelines. Karolin Frank's study offers an important perspective of linking the historic preservation movement to larger questions of city planning. As the fortunes of cities like Boston, Philadelphia, Charleston and Savannah rebound due to their rich – and surviving – architectural heritage and humane sense of scale and quality inherited from the past, other cities will be required to take notice. This study will help them discern just why it is those cities succeed.

Robin B. Williams, Ph.D.
Chair, Department of Architectural History
Savannah College of Art and Design
Savannah, Georgia

List of Chapters

Contents

List of Figures

List of Tables

List of Maps

List of Photographs

Abbreviations

AIA	American Institute for Architects
ACHP	Advisory Council on Historic Preservation
ARB	Architectural Review Board (also: BAR)
BAR	Board of Architectural Review (also: ARB)
BHAC	Beacon Hill Architectural Committe
BHCA	Beacon Hill Civic Association
BLC	Boston Landmark Commission
Br.	Brochure
BRA	Boston Redevelopment Authority
BSA	Bostonian Society of Architects
CBD	Central Business District
CC	Center City, District in Philadelphia (equals the CBD)
CCD	Center City District
CDBG	Community Development Block Grant
CPS	Current Population Survey
DAR	Daughters of the American Revolution
Dept.	Department
DVRPC	Delaware Valley Regional Planning Commission (founded 1965)
EU	European Union (since 1.1.1994)
GMP	General Master Plan
HABS	Historic American Building Survey (founded 1933)
HABSI Form	Historic American Building Survey Inventory Form
HAER	Historic American Engineering Survey
HBI	Historic Boston Incorporated
HCF	Historic Charleston Foundation
HSF	Historic Savannah Foundation
HUD	Department of Housing and Urban Development
ICCROM	International Centre for the Study of the Preservation and Restoration of Cultural Property (Rom)
ICOMOS	International Council on Monuments and Sites (Paris)

ITC	Investment Tax Credit
Iv-"City"-"number"	Interview (see Appendix A.1)
MSA	Metropolitan Statistical Area
NDP	Neighborhood Development Plan
NHLP	National Historic Land Program
NHPA	National Historic Preservation Act of 1966
NIMBY	"Not in my backyard"
NPS	National Park Service
NR	National Register of Historic Places
NSDAR	National Society of the Daughters of the American Revolution
NTHP	National Trust for Historic Preservation
OPDC	Old Philadelphia Development Corporation
PHC	Philadelphia Historical Commission
PHIL PLAN	Publication of the Philadelphia City Planning Commission
PHL	Philadelphia, Pennsylvania (consolidated city boundaries from 1854)
PCDC	Philadelphia Commercial Development Corporation
PECO	Philadelphia Electric Company
PIC	Private Industry Council (Philadelphia)
PIDC	Philadelphia Industry Development Corporation
PPD	Print and Picture Department (in the Free Library of Philadelphia)
PPP	Public Private Partnership
PSMA	Primary Standard Metropolitan Area
SHPO	State Historic Preservation Officer
SMSA	Standard Metropolitan Statistical Area
THPO	Tribal Historic Preservation Officer
TPL	Trust for Public Land (founded 1979 by the NationalTrust for Historic Preservation)
RDA	Redevelopment Authority (of the City of Philadelphia)
UN	United Nations
UNESCO	UN Education and Scientific Cultural Organization
U.S.(A.)	United States (of America)
WHIN	World Heritage Information Network
ZBA	Zoning Board Adjustment

1 Introduction

1.1 Thematic Introduction

The some 500 years that have passed since the discovery of the American continent represent a relatively short period of time compared to the long history of European settlement; nevertheless, the people of the United States has at its disposal a significant number of historic buildings and ensembles constructed between the 16th and 19th centuries. In the cities of the thirteen founding states and the south-eastern coastal states alone there exists today a remarkably large number of historic buildings which are indeed worthy of an adequate preservation program. The restoration, reconstruction, conservation and conversion, as well as the sustainable maintenance of these historic buildings must be considered an important part of the preservation of American architectural and cultural heritage. The object of this book is both to identify and analyze the nature and the extent of U.S. building and monument protection schemes, as well as the motivation of the different protagonists behind such schemes.

Since the 1960s, public attention has been drawn increasingly towards the thematic link between the historic preservation movement and urban planning. Nowadays, the organized preservation of buildings and monuments is more than a mere "yearning for history"; it represents an active and integral part of urban planning in U.S. cities. Whilst studying the preservation of historic buildings on an urban level, one soon encounters the concept of "historic districts"[2], which has particularly area-specific as well as general characteristics. In order to approach these involved planning, economic and social issues, this study will focus analytically on the four historic districts of four cities. In order to provide a comparative regional overview, those chosen are the cities of Philadelphia, Boston, Charleston and Savannah, all of which were historically significant U.S. East Coast ports and lay within the former English colonial territory (see Map 1.1).

---

[2] This central term, "historic district", is impossible to translate within the German context. German concepts of *Schutzbereiche*, *Schutzgebiet*, *Ensembleschutz*, *historischer Bezirk* or *flächenbezogener Denkmalschutz* do not correspond with the U.S. concept of historic districts because they already have definitions specific to the German context (see: Precht von Taboritzki 1996, pp. 48). The main differences can be identified in the motives behind the establishment of such areas and their administrative elements (guidelines, legal status, degree of preservation).

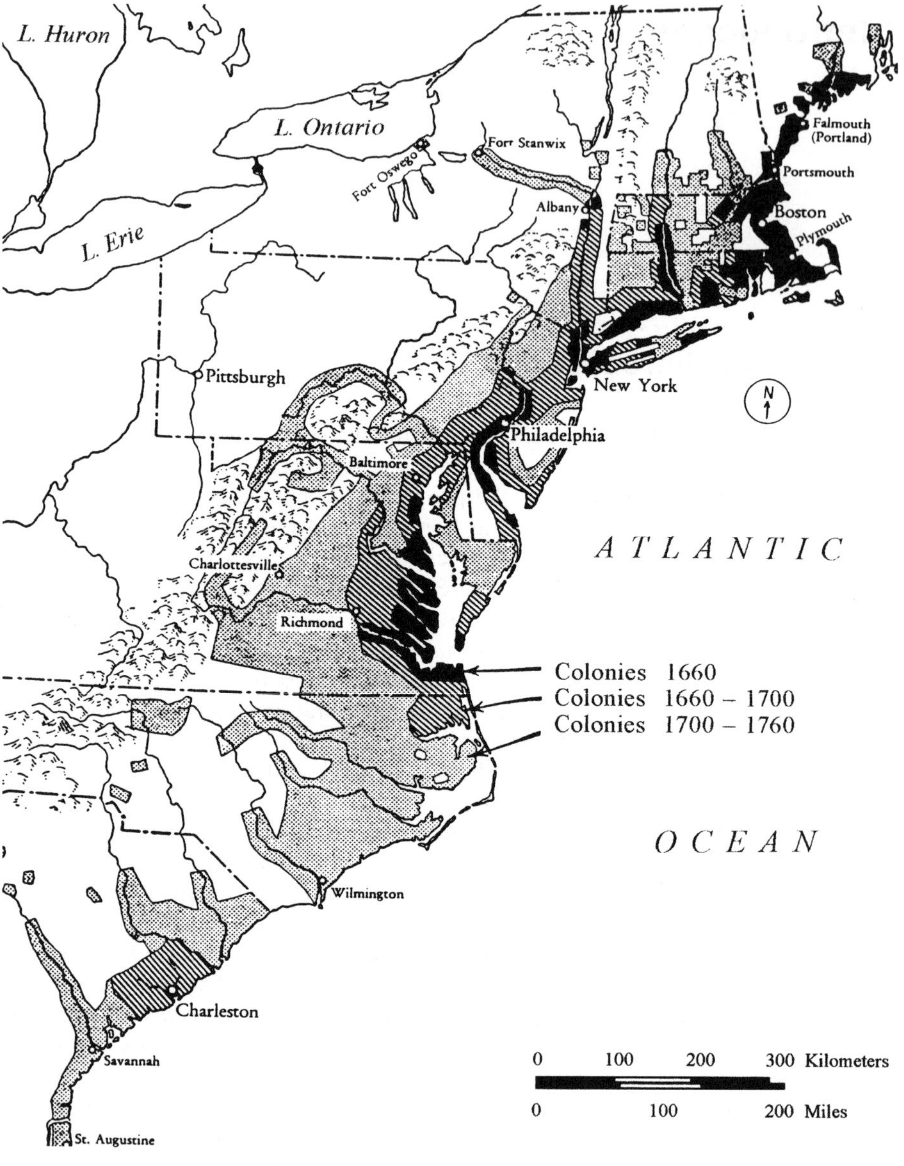

Map 1.1. Extent of English colonial territory from 1660 to 1760 (Source: Dahms 1983, Appendix; modified)

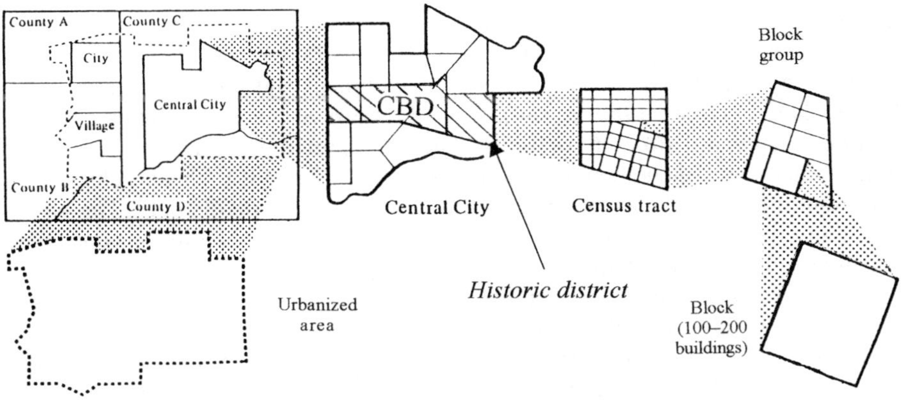

Figure 1.1. Schematic location of the four historic districts within the U.S. urban system (Source: Bourne 1981, p. 65; modified)

All historic districts examined in this work are to be found within the demarcations of old inner city areas and belong to the modern-day Central Business Districts (CBD). The selected cities are of national importance, though they play no special role within the U.S.A.[3] On closer examination, the American Civil War (1861–1865) shows us an interesting contrast between the northern cities (Philadelphia and Boston) and the southern cities (Charleston and Savannah), which leads us to accord particular significance to these elements within the analytical part of this study.

By applying a variety of interdisciplinary methods, this work intends to show that the preservation of monuments within these historic districts throws a new light on the dynamics at work within the urban areas of the United States.[4] In the process, the planning, economic and social complexity of this concept will be particularly stressed.

1.2 Scientific Context

This examination of historic districts within the field of geography will be situated specifically within the scientific context of urban geography. No single scientific system exists for the study of urban geography, which instead involves various fields of human science and technology in order to understand the nature of the physical environment (Lichtenberger 1998, p. 18). These related fields interweave to avoid the formulation of one-sided hypotheses and, besides enabling these

3 Unlike, for example, the capital, Washington, DC, the "melting-pot" of New York City or the Los Angeles cityscape.

4 See also Datel and Dingemans 1980.

fields to work towards a common hypothesis, help to reach superior conclusions, as well as providing key information on regional processes. Through the collaboration of planners and scientists with interdisciplinary training, many misunderstandings and misinterpretations may be avoided. Korte indicates the contradiction inherent in cities between fixity (urban development and architecture) and flux (social change) (Korte 1989, p. 184). It would, however, be too complex a task to examine the overall structure of a city. Thus it is preferable to favor an "analysis of the general city phenomenon" (Lichtenberger 1986, p. 388). This is achieved by selection of a particular urban area, in this case, the "historic district". The physiognomic and cultural dimensions evident within these areas will be considered and assessed from all standpoints.

The preservation of historic buildings is classified as an interdisciplinary research area within the field of urban geography. It is made up of the scientific fields of urban geography, urban development, architecture and social history (Lichtenberger 1998, p. 25). Taking into account the available instances of collaboration between urban planning and the preservation of historic buidlings, the following may be observed (Figure 1.3).

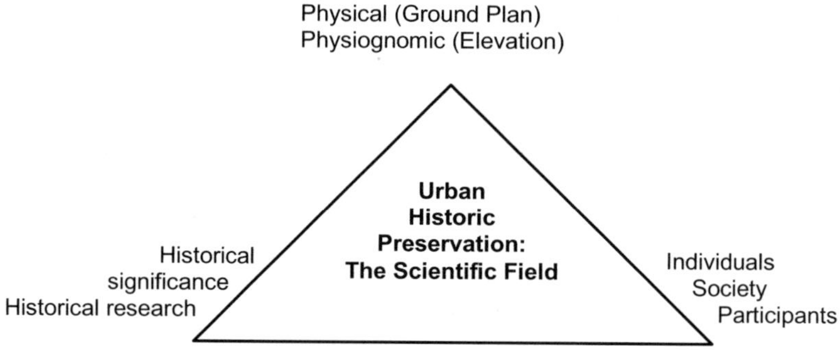

Figure 1.2. The dimensions of historic preservation

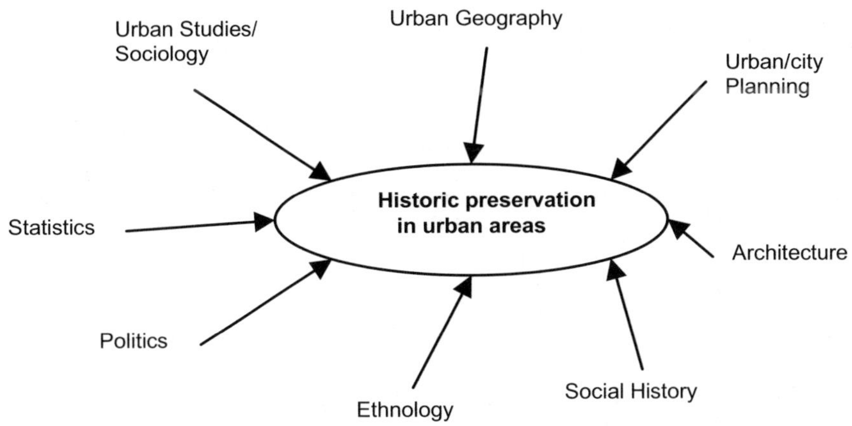

Figure 1.3. Combination of various specialist fields within the field of urban historic preservation (Based on: Lichtenberger 1998, p. 25)

1.3 Main Areas of Research and Object of this Study

Which aims, motivating factors and aspects of urban planning play a significant role within the historic districts of the United States will form the basis of this research study. The presence of varying ethnic, generational and social groups is reflected within American cities and shows the great number of conflicts of interest and area-related conflicts that exist. These social and residential needs involve a wide range of social identities and planning parameters, which can also be identified within the historic districts. Rather than seeking to focus on a fictitious urban society, this study will aim to analyze the concrete motivations and interests at play, using a variety of analytical approaches (Chapters 5 to 8), as the shaping and development of urban areas is determined not only by physical but also by social considerations. It is not the object of this work to undertake a retrogressive study of the transformation of the process of historic preservation. Instead, its aim is to concentrate on the practice-orientated planning of historic districts and the economic implications for the cities' future (how the city will subsequently develop). As a result of this analysis, answers to the following question relating to the historic districts of the four cities of Philadelphia, Boston, Charleston and Savannah should emerge:

How, in historic districts, may culture and history be linked simultaneously within the modern urban environment and the social and economic flux of a city, independent of time and yet taking into account the passage of time?

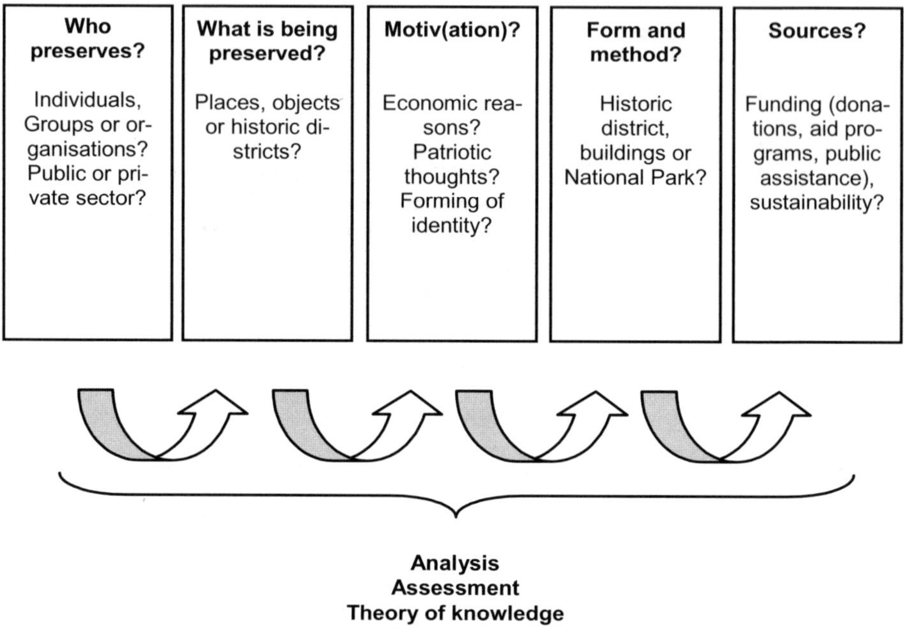

Figure 1.4. Historic preservation: fundamental theoretical questions

Many different theories and concepts play a prominent role here: historic districts can, on the one hand, be created by urban planners as a means of preventing homogenous and thus exclusive groups settling in one particular area. On the other hand, the creation of historic districts may also be viewed as a political tool, intended to give form and direction to urban development and thus enable the authorities to work towards achieving better integration, on both a social and an economic level. Figure 1.4 (below) shows more clearly the questions which form the basis for this study.

Figure 1.4 shows clearly that the motivations and the financial support behind the preservation of buildings, as well as the form and extent of the preservation, are extremely complex issues which must be taken into account when an historic district is created. The various parameters can be linked in a number of ways.

1.4 Structure of the Analysis

The aim of the first part of this study is to introduce the reader to the highly specific terminology involved. At the same time, the data and information sources of the research subject and the areas of analysis will be explained in detail (Chapters

1 and 2). After a consideration of current research theories on this subject, there follows a theoretical introduction, focusing on the origins of the desire to preserve monuments and buildings of interest. A brief explanation of the nature of U.S. urban areas will also be provided (Chapter 3). Finally, taking into account important urban parameters, we will analyze just how far the establishment of an historic district can create a further dimension within the field of urban planning. After the theoretical focus of the preceding chapters, Chapter 4 will concentrate in part on practice and present the various private and national organizations which work towards the preservation of historic buildings, nationally, within a single state and locally. The reasons and methods behind the preservation of these monuments, as well as the underlying concepts and influencing factors, play a particularly significant role here; they are judged on three levels, forming the basis for the analytical element of this study. The theoretical conclusions of the preceding chapters and the three levels of judgment in Chapter 4 will then be used in the analytical part of this study, concentrating on specific historic districts, namely those in Philadelphia, Pennsylvania (Chapter 5), Boston, Massachusetts (Chapter 6), Charleston, South Carolina and Savannah, Georgia (Chapter 7). Those findings and results of general and comparable significance will then be discussed in Chapter 8 within an U.S. and German context.

1.5 Starting Point and State of Current Research

In the following section, our aim is to provide a brief overview of the current state of research within the field of geography as far as the preservation of historic buidlings and urban planning are concerned. We will differentiate between sources published in the U.S. and those published in Germany.

A wide variety of publications have appeared over many years in the U.S.A. focusing on the subjects of historic preservation and urban planning. It is rare, however, that such studies seek to link these two subjects (Birch and Roby 1984, Cullingworth 1997, Murtagh 1997, Hodder 1999). This is due, on the one hand, to the difficulty in linking two separate urban research areas and, on the other, to a lack of previous analytical research in this area.

The available publications may be divided chronologically and thematically into six areas:

1. From the beginning of the 20th century to the outbreak of the First World War most publications were by private individuals who concentrated exclusively on one object and whose plans and suggestions were intended to ensure preservation alone (Hunter 1991, NSDAR 1979).
2. Around 1900, the City Beautiful Movement came into being. In connection with this movement, several studies were published which proposed a system to protect certain areas. In order to achieve this, a more extensive planning process was required. We may take as examples the contributions made by Peterson

(in: Krueckeberg 1997, p. 40), Wolf and Wilson (in: Campbell and Fainstein 1996, p. 68).

3. From 1950 onwards, questions began to arise as to the social and economic implications of cities, due mainly to the massive suburban developments taking place in American cities. The question of the preservation of historic buidlings was only discussed peripherally. However, of interest to researchers is the fact that in no instance did developers make any attempt to take into account the historical building materials when working towards urban renewal in city centers and historic districts[5] (RDA 1968, Murtagh 1997, p. 62).
4. After 1960, some projects were set up to examine the aims and suggestions of some city planning boards. These aims were applied to wider areas and were controlled primarily by the individual states (Murtagh 1997, Venturi et al. 1997).
5. Towards the end of the 1970s, writers began to consider the economic implications of the preservation of historic buidlings, and to analyze its potential (Preservation Alliance 1997b, NTHP 1975, NTHP 1998b).
6. Over the last few years, the tourism industry has played a significant role and is at the heart of the debate in which cities are now seen to occupy a place within the global urban network whilst maintaining their individual characteristics. Taxation, the preservation of monuments, education[6] and questions of heritage[7] have assumed particular significance within the area of preservation (Heritage Foundation 1985, Katz 1994, Ellin 1996).

These six areas show that the general public's interest in architecture and history has changed dramatically over the last few years. Amongst other factors, the Civil Rights legislation of 1964/65 may be seen to have initiated a reawakening of consciousness in the American people.

In 1980, Datel and Dingemans undertook a study on the historic preservation, focusing on urban and social change. In 1984, Birch and Roby published a groundbreaking article in which they investigated the conflicting and simultaneously interdependent areas of urban planning and preservation of historic buildings. Lowenthal (1988) made a significant contribution with his study on regional history. Fitch (1995), a professor at the University of Pennsylvania in the Historic Preservation department, published, in 1990, an extensive textbook on the fundamental theoretical principles of the preservation of historic buildings, the first of its kind. Murtagh (1997) published a monograph, in which he brought together and examined the various practical disciplines required for the historic preservation. In 1997, Cullingworth published an extensive study, entitled "Planning in the U.S.A.". In this work, he looks at the written legislation, the motivations at work

[5] Such districts were, in fact, little more than areas of land containing a significant number of features of historical interest but with no defined administrative system.

[6] Education refers to the concept of enlightenment as well as the awareness of historic preservation.

[7] In this study, heritage refers to the cultural and architectural "inheritance" of a particular group within U.S. society. Chapter 3 provides a specialist, methodological discussion of terminology.

and the processes that take place in the field of urban planning in the United States. In Chapter 9, when discussing "Historic Preservation", he approaches the question of the preservation of historic buildings by focusing on selected regional examples (Cullingworth 1997, pp. 113).

As well as these authors and collaborative works which concentrate exclusively on historic preservation, there are many publications which deal with related theories and methods of historic preservation. At this point it is essential to mention Jacobs' classic 1961 publication "Death and Life of Great American Cities" (Jacobs 1993). In her monograph, the author describes, amongst other things, the decline of historically important neighborhoods due to a lack of respect and a destructive urban planning policy, evident within the cities as a result of the Urban Renewal Projects[8]. Jacob's interest, however, lies primarily in the consequences for the urban social and economic structures, which, in the context of the time at which she was writing, means that Jacobs would not have suggested historic preservation as an integral part of urban planning.

Aesthetic concerns in the field of (historical) urban history were integrated into research as a result of Venturi, Brown and Izenour's "Learning from Las Vegas" (Venturi et al. 1997[9]). They first criticized the wild developments in American architecture in 1972 and thus founded the basis for the Urban Design Movement[10]. Lynch (1997), who had studied under Frank Lloyd Wright, attempted to concretize theories which would come closer to setting up a universal standard by which "good cities" might be judged. At the same time, he takes into account the many conflicts of interest, the land development laws, the economy and the vast number of social groups living in the metropolis. In a subsequent publication (Lynch 1982), he looks more closely at the social image of the city and factors such as acceptance and identity within the urban social environment. In this study, historic preservation is an important element in the preservation of aesthetic urban characteristics and the attention given to promoting a favorable image of the city. In this context, Kunstler (1994) criticizes the fact that this yielded little change in urban areas. In some cases, the suburbs of American cities, being richer than their inner-city counterparts, exist today entirely independently of their mother city.[11]

German authors have rarely focused on the question of urban historic preservation in the United States. Bornheim (1956) was the first to publish on this subject

[8] Public programs that are financially supported and organized through the Urban Renewal Program of the U.S. Department of Housing and Urban Development (HUD). These programs were only concerned with the renewal of existing structures and not society-related neighborhood revitalization.

[9] This book was first published in 1977.

[10] Research into individuals' perception of their environment identifies a growing tendency to value aesthetics and a greater emotional awareness of buildings and urban areas. The Urban Design Movement laid the basis for specialist discussions on this theme (see also Garvin 1996).

[11] Kunstler refers to this idea in his 1998 monograph, in which he includes a series of examples of projects and methods by which cities may undergo a positive process of revitalization.

with an article on historic preservation in North America. One theory is that European prejudices against U.S. cultural heritage programs have been extremely detrimental to research and the exchange of information. Many trends described by Bornheim are still in evidence today. In the U.S., historic preservation is more a general concern which focuses on specific areas of particular interest (historic districts) on a national and regional level. Through Bornheim, many others identified insufficiencies in the system of historic preservation (a lack of preservation laws and financial support), which have since been recognized and rectified. In 1977, Uhlig published a book entitled *Vereinigte Staaten von Amerika: Zur zukunftsorientierten Bewahrung erhaltenswerter Städte* ("United States of America: The Need For a Forward-looking Policy of Protection for Historically Important Cities") as part of a series of studies on urban preservation. According to Uhlig (1977), historic preservation in the historic districts works as a catalyst and not as a planning method aimed at revitalizing the district. Hofmeister's essay, *Die Erhaltung historisch wertvoller Bausubstanz in den Städten der U.S.A.* ("The Preservation of Historically Important Structures in U.S. Cities"), also appeared in 1977. He saw the American bicentennial as a particularly important motivation towards the better preservation of historic buidlings. Hofmeister's article lays a basis for historic preservation in the United States and his regional studies will be discussed in more detail in the analytical part of this work.[12] More recent authors include Born (1995), who examines the regional significance of the exchanges that take place between the various protagonists who cooperate to preserve cultural heritage (*Historische Kulturlandschaft*[13]) in Germany and New England. Similarly, Born indicates that these exchanges suggest that historical and geographical concerns are not necessarily the strongest motives at work in the field of historic preservation. Above all, his studies show the social, legal and administrative differences between the two geographic areas.

Clearly, there are few studies to date which examine the individual administrative, planning-related and informal processes and methods within the urban historic districts. Moreover, few analytical studies have been undertaken in the historic districts which look in detail at the economic and social implications of the relevant program of historic preservation. These gaps must be filled in the study which follows.

[12] Particularly in relation to Savannah (Hofmeister 1977, pp. 138 and pp. 146) and Charleston, SC (Hofmeister 1977, pp. 148).

[13] Born (1995) deliberately capitalizes this term in order to stress the historical qualities of cultural heritage.

2 Methodology and Terminology

2.1 Sources

Macro-level		Meso-level		Micro-level
National (i. e. tax deductions)	→	States	→	City Council
National Trust, U.S. Dept. of the Interior	→	Local/regional/ national groups, Preservation Alliances	→	Neighborhood organizations, Active individuals
Society (i. e. tradition of capitalism)	→	Specific urban conditions (social, ethnic mix)	→	Subjective perceptions of Individuals within a specific area, quality-of-life-issues[a]
Whole structure of the USA (existing investigation: National Register)	→	Historic districts, area under special protection, cities	→	Specific buildings, museums, research of a certain time frame
Economics and historic preservation	→	Research of existing potential: What can be used?	→	Touristic benefit, re-use of old buildings

[a] Quality-of-life-issues include, for example, the absence of criminality, vandalism and/or noise pollution. These are often a high priority of neighborhood organizations.

Figure 2.1. Various levels of research in the field of historic preservation

In order to analyze the four historic districts and their organizational structures, related fields of interest and potential, we must first of all examine the theoretical and administrative basis upon which U.S. historic preservation programs are founded. This analysis must, in turn, be divided and differentiated into the three fields of micro-, meso- and macroanalysis, which will then allow us to form observations on the regions in question.
According to the criteria chosen, it is possible to identify the conflicts of interest involved in the historic preservation programs (land use, economic issues, urban image and structures, how the city is perceived from within and externally).

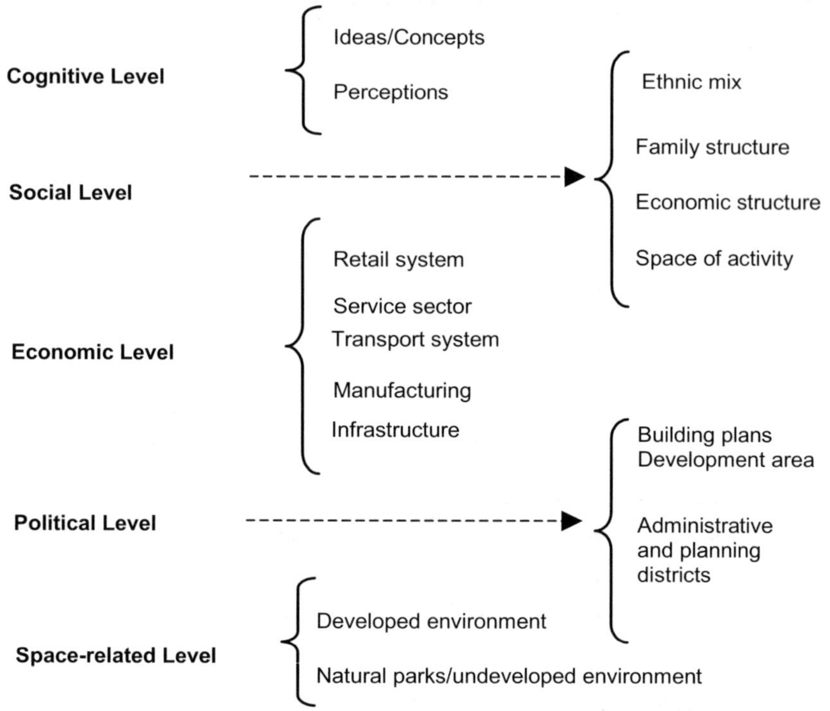

Figure 2.2. Research categories in the fields of historic preservation and urban planning (Based on Ott 1999; following Murdie 1969)

The multiplicity of methodologies consists of primary research, including specialist interviews, the creation of regional maps and periods of regionally-specific observation, as well as secondary analysis. This combination is necessary in order to achieve a complete overview of the historic districts.

Specialist and intensive interviewing took place in all four cities.[14] The interviews in question were non-standardized and topic-related (for a complete list of interviews see Appendix A.1). Various official representatives and specialists were interviewed, some of whom were involved officially and some privately in urban planning and/or historic preservation programs. The interviews were used to form the basis for secondary analyses, draw substantial conclusions and to support arguments and individual observations. These may be found in Appendix A.2 as part of the interview-related topics. Moreover, an integral part of the research process was the attendance, in the capacity of observer, of hearings and meetings[15] organized by the local authorities or institutions, in order to collect further primary

[14] Interviews can be identified throughout the text by the following abbreviation: "Iv-City-Number".

[15] For example, the monthly Architectural Board Meetings of the Philadelphia Historical Commission (PHC).

source material. Some of these gatherings were open to the public (public hearings), others only to a certain extent (restricted hearings). In other words, the latter could only be attended on receipt of an invitation or a summons. Contrary to the interviews, it was possible here to gather authentic information. Although the author's primary analysis is given priority in this study, other sources were also taken into account.

The secondary sources are made up of statistical information and planning documentation. Accounts of social distribution within the region and local regulations regarding historic preservation are of particular significance (historic district ordinance, historic zoning ordinance). Furthermore, we will draw on the results of the U.S. Census, as well as the publications of national, state and local authorities and the individual organizations responsible for historic preservation (for example, the NTHP (National Trust for Historic Preservation) or the AIA (American Institute for Architects). The structural organization of the neighborhood organizations or other institutions provide further source material in which results have to some extent been assessed and catalogued in graph form. Finally, construction maps and plans, specialist newspapers, statutes and the local press complete this list of source materials. The secondary analysis consists mainly of structural data, whilst the primary sources also enable us to assess the exchanges that take place within the historic districts.

The combination of theory and analysis in this study represents an inductive process which looks in detail at the various elements and seeks to differentiate between the results obtained in order, finally, to reach conclusions that can be applied in totality to the urban historic districts. The statistical overview of a neighborhood is for this reason only used as a example in the case of the historic district of Society Hill in Philadelphia.[16]

2.2 Terminology

It is necessary, when introducing terminology, that the specialist vocabulary, categories and foreign terms be fully explained. This is particularly important to explain terms whose meaning varies according to the cultural environment in which they are to be found. Such is the case with terminology from British, American and German contexts.

In order to represent fairly the different ethnic groups living in the U.S., the American genealogical terminology will be applied, including, for example, the terms "African American", "Hispanic" and "Native American". Our aim is to respect the ethnic definitions used by the U.S. Census since 1980 and which were chosen by the groups in question as representative of their identity. The African Americans will figure in this study on several occasions as an ethnic minority. Thus, the examination of relevant structural data applying to this group in the historic districts may subsequently be compared with the results obtained for white

[16] The full name is: Society Hill and Pennsylvania Hospital of Washington Square West.

Americans. The statistical differences are particularly clear between the latter and the African-American ethnic minority.[17]

It is especially problematic to define the term "neighborhood" (see also Weiler 1974 and Behrens 1995).[18] This influential social community is a particularly unique American form of organization and is not always defined by administrative or geographical borders (districts). Even the borders of the census tracts, whose demarcations are reassessed every ten years, are not always identical to the neighborhood borders.[19]

Available literature provides information on the main dates and statistics of various eras of American history. However, much numerical information has been lost which did not correspond to the prevailing opinions and practices of the time (see Appendix, Figure A.3).[20] The definition of the term "historical" will thus be assessed according to the existing laws and statutes of each individual city (see table 4.2 in Chapter 4 for comparison).

This study is not intended to be clearly gender-specific. The author has chosen to adopt the masculine form and will thus refer to an urban planner, for example, as "he" and not "he/she", as a form of shorthand intended to aid the reader. Unless specified, the masculine form refers to both sexes. Finally, the glossary in Chapter 9 provides further information on and definitions of terms and institutions mentioned in this work.

---

[17] A discussion of the historical, social and biological concept of "race" can be found in Alba's work (1990, pp. 10), for example.

[18] A literal translation into German (*Nachbarschaft*) fails to reflect the true meaning of the word.

[19] Nor is the word "suburb" translatable into German. The term *Vorort* in German refers to populated areas lying outside the city itself, whilst in the United States, whole areas of a city, such as West Philadelphia, for example, are designated as suburbs, even though, since its construction, the area has always fallen within the city limits. This vague formulation of the term "suburb" is discussed in more detail by Jackson (1985, p. 4) in his monograph.

[20] See in particular the works of the American Foundation for Architecture (1991), Whiffen and Koeper (1981), Whiffen (1996) and Walker (1997) (see Appendix, Figure A.3).

3 Historic Preservation and Urban Planning

3.1 Historical Aspects of Preservation Programs

In order to write on the subject of historic preservation, it is first of all necessary to explain the basic historical elements involved as well as the meaning and semantics of the word. The questions to be asked in a research study which seeks to interpret history and historical sources are the same as those to be asked when considering (historically significant) buildings:

1. What does the documentation suggest and is it selective?[21]
2. Why are certain aspects present and what elements are missing?
3. What picture of reality is presented to us?

Terms such as "heritage" and "historic" are encountered with increasing frequency in the United States. Perhaps this is due to a sense of uncertainty with regard to all that concerns the future, and the terms allow society to look backwards rather than forwards. It is so difficult to keep track of all that modern life brings with it that the clarity of history, with its fixed demarcations and documentation, more easily satisfies the general need for a sense of identity. Some critics go so far as to state that, as far as Germany and the rest of Europe are concerned, history has been defined and concretized in such a way as to combat the current pessimistic attitude towards the future. As historic preservation within the historic districts is now becoming an issue within the field of urban planning, this belief in culture can be concretized and even manipulated.

In his book *"Das kulturelle Gedächtnis"* ("Cultural Memory"), Assmann (1999) puts forward the theory that only taking into account the whole of human experience can lead to an overall picture of reality. This is too complete a requirement to be realizable, as experience is infinite, and we are obliged from the outset to be selective. Why, then, is memory so important to us? Evidently, a human need exists to make past thoughts and decisions, as well as their motivations, accessible and obvious to future generations. Gutschow (1978, p. 131) describes this need as a *Sehnsucht* (yearning).[22]

[21] Due to the World War II, there is of necessity less choice of documentation available in Germany. This is more a question of circumstances than of choice.

[22] In his monograph "Icons and Aliens", Costonis (1989) gives concrete examples of the demolition myth and reconstruction projects aiming to remain true to original structural features.

Historic preservation and, thus, the preservation of historic districts can be seen as a reflection of this desire. It is widely accepted that the 18th century world of Goethe felt no need to base itself on traditions and fixed social systems.[23] Today, the ideas of Goethe have probably only a relative significance. The Americans have themselves realized – following the European example – that assigning value to the past and cultural heritage can open up new perspectives and may have social consequences (Fitch 1995, p. 23). According to Fitch, the maintenance of historical heritage and historic preservation are currently "modern" concerns in the U.S.A. (ibid. p. 41). Evidently, historic preservation gained particular significance during and after the U.S. bicentennial. The Declaration of Independence, and not the colonial era, led to a new awareness of history. The history surrounding the First Americans, or Native Americans[24], is virtually ignored (Iv-Phl-3[25]).

Determining what is objective and subjective and thus, to a certain extent, what is true and what is false, is a difficult task for historians, as sources and data are all too often incomplete and consequently open to a broad range of interpretations.[26] It is almost impossible, in the end, to decide what motivations lay behind historical events, and conclusions are, more often than not, merely speculations. These factors mean that historical research may draw interpretative conclusions and subjective elements cannot be avoided. In many cases, historical researchers in the field of architecture have at their disposal only one picture or plan, drawn from a single perspective, in order to reconstruct the structure of a particular building. It is rare that these sources also provide information as to the building's historical and cultural significance and this, in turn, affects the interpretation. In some cases, where the archives provide little guidance, historical authenticity can only be relative. The aim is often more to seek a cultural standpoint and define historical events from a social perspective than to provide historical truths. It is an all too fine dividing line that separates historical and contemporary accounts of "fact".[27]

Besides this cultural perspective on history and the past in an urban context, there are also economic factors which must be taken into consideration. Thus, Costonis (1974) writes that history and the past must be considered as an integral part of the urban planning process, as this would constitute an effective means of securing financial support for historical buildings and features.

To what extent buildings and other structures may be considered culturally meaningful is a question that is open to individual interpretation. We must also ask ourselves whether certain structures might be considered less worthy of attention

---

[23] See the poem by Goethe cited at the beginning of this book.

[24] It can be assumed here that it is particularly difficult to integrate Native American history into urban planning projects as very few physical structures exist to represent this population group.

[25] This abbreviation refers to the specialist interview carried out (see Appendix A.1).

[26] In his monograph, "Writing History", Veyne (1984) discusses the nature of history and points to the inevitable manipulation of fact that of necessity occurs in the process of interpreting and recording history.

[27] The Betsy Ross House in Philadelphia is a good example of this: Betsy Ross (allegedly) designed the first U.S. flag with its striking 13-star motif in this house.

because they recall unhappy eras. The historical value of a building can also depend on how efficiently it has been maintained, which can affect its original significance. This discrepancy between esoteric (scientific significance and historical value) and exoteric ("open to all") must be recognized and dealt with appropriately by all involved in the process of historic preservation. "Culture" is thus concretized through functional mental processes and these allow the building to be elevated to the status of a monument (Heinz 1991, p. 305). In order to include all ethnic minorities in this study, we must first of all identify the particular group to which they belong.

Architecture must be seen as the art of necessity, inspired as it is by practical needs. A building must thus have a function in the modern world and this is a criterion, in the field of historic preservation, that must be considered alongside the assessment of its individual worth. What form this assessment takes and who is responsible for it plays a vital role (Kiesow 1995, pp. 88)[28]. Scully (1988), however, contests this view, stating that no difference should exist between architecture and urban planning. A building should therefore not only be acceptable as an individual object, it should also fit into the architectural context of the city.

A study published in 1980 by the National Trust for Historic Preservation (NTHP) concluded that, by maintaining historically important structures, neglected historic inner-city areas constituting the economic nucleus of the city could be revitalized. As a result, the National Main Street Center was opened in order to help restore the economic importance of these historic commercial districts through historic preservation programs. The idea was to assure low rents and purchase prices (Ziegler 1971, Matthews 1992) and to enable neighborhoods that make up the district to be redeveloped and redefined. President Jimmy Carter initiated a series of studies in the 1980s which proved that there was a clear link between the preservation of historic structures and economic success.

It is surprising how many institutions in the United States play a merely advisory role in the field of historic preservation. Executive associations figuring in this study are mainly drawn from the city halls, which indicates that historic preservation work is primarily a local political issue. Which priorities and motivations underpin historic preservation programs and how they are apparent will be analyzed in more detail in the more practice-orientated Chapter 4 of this study.

Historic preservation can also be used, in political and economic terms, as a means to achieve power. According to how public and private areas are preserved, it is possible to control the culture, society and area to a large extent. Moreover, the fundamental question always recurs: How authentic must, or should, the preservation work be? Whatever criteria are proposed to regulate the elements of control and consequence, in the end the local authorities make the decisions.

---

[28] Further historic preservation and urban planning issues within a German context can be found in Chapter 4 of Kiesow's book (1995) *"Denkmalschutz als Teil des Städtebaus"* ("Historic Preservation as a part of urban planning").

3.2 The Term "Historic Preservation" and its Semantics

Table 3.1. Semantic connotations of the term "historic preservation"

USA	EU (except UK)	United Kingdom
Historic preservation (*Bewahren*)	Protection (*Schutz*)	Conservation (*Erhalten*)

Source: based on Fitch (1995, p. 39)

Much like the German term *umweltfreundlich* (environment-friendly), there are no official definitions for the terms "historic preservation", "heritage" and "landmark"[29]. Consequently, these terms must be explained in context in order to enable an objective analysis of the processes of historic preservation.

The creation of protected areas involves the establishment of higher authoritative controls or laws which are usually applied locally by different institutions. These four categories have been implemented in historic districts in order to define these four controls (from: APA 1992, p. 17):

1. New construction
2. Demolition
3. Alterations
4. Additions

Nowadays, historical city areas in the USA are no longer at the economic heart of the inner city. Further factors include the fact that many structures are often in poor condition and the historic districts have suffered both socially and physically. Many architects and urban planners believe that historical continuity, structural identity, revaluation and revitalization should be respected in order to enable the historic preservation program to be carried out effectively. It is, however, often expensive to maintain older structures and American individualists are never pleased to see limitations and conditions imposed on how they choose to go about their business. Many investors are put off by the size of some of the monumental factory buildings, but also in part by the restoration and restoration work that they would be obliged to carry out in order to clean up the detritus left by past generations.

Under such circumstances, can historic preservation be incorporated and accepted within the system of private capitalism? According to Fitch (1995), historic preservation should be organized in such a way as to lead to significant economic advantages in the long term and simultaneously revitalize the area. He refers particularly to the tourism industry, general revaluation of property and new areas of commercial importance (internet and advanced technology). The economic gain is thus dependent on the financial support provided and the execution of historic preservation work.

[29] For further references, see also glossary.

Historic preservation work is as much a public as a private undertaking, even if unwillingly executed on occasions. From a technical point of view, Feilden (1997 pp. 7) is able to differentiate between seven forms of intervention involved in the preservation of historic structures. These forms of intervention will be considered again in the analytical part of this study:

1. Prevention of structural decay
2. Maintenance of the current condition
3. Maintenance of the current structure
4. Restoration and partial reconstruction
5. Rehabilitation (restoration)
6. Reproduction
7. Reconstruction

Tunnard (1975, pp. 87) points out that 70% of historic preservation work in the United States is carried out by private individuals and institutions. This situation has, to the present day, changed very little (National Park Service et al. 1994). Hence, there are countless private organizations, societies and associations, all with different interests at heart, whose aim it is to carry out historic preservation work and protect historical and cultural heritage.

3.3 Heritage as an Element of Historic Preservation

As "heritage" is a term that recurs frequently in connection with historic preservation, it should firstly be explained in more detail. This presents a number of problems due to its multiple and, to some extent, indirect usages.

3.3.1 Heritage and Genealogy

Due to genealogical research in the United States, the term "heritage" has come to imply elements of origin and ethnicity. Heritage research is a means of assigning an individual identity to the ethnic minorities that have developed within the culturally pluralistic U.S.A. In this context, culture and heritage are clearly interdependent. Whenever heritage is the focus of a publication, it is usually in relation to social or ethnic groups. Because so many ethnicities converge in the United States, Hodder (1999, p. 439) describes heritage as an urban resource. The heritage of a particular social group is an important factor within the global and post-industrial national economy (ibid. p. 439).

It must be noted that the Native Americans, whose earlier presence suggests that American history should in fact begin with them, possess no architectural structures. Hence, their historical inheritance consists mostly of historic sites and the continuation of tradition through presentations and the sale of hand-crafted items made by modern-day generations. Another significant minority are the African Americans who, in recent years, have affirmed their own history through "Black Heritage".

The awareness of and emphasis on cultural heritage, as well as genealogical research into the question of origin, is a relatively new phenomenon. There is, however, evidence of social resistance to this phenomenon, which protests against the many individual programs that support ethnic minorities. Today, groups of European origin are also differentiated socially within the United States (Irish Americans, Italian Americans); this social phenomenon discussed by Alba in his 1990 work.[30] It may be observed that Americans' growing awareness of and emphasis on heritage is above all a consequence of the Civil Rights Act of 1964/65. The questions of belonging and identity, as well as individual self-evaluation within U.S. society, also play an important role. Historic preservation is the cultural and regional reflection of this growing awareness and heritage movement.

The inclusion of ethnic minorities into the historic preservation programs is certainly a reason why this aspect of heritage has been promoted and supported by state authorities. We might ask whether it is in everybody's interest that minorities play such an influential role in the heritage-led preservation programs, as many of their so-called "urban experiences" fail to reflect positively on the U.S.A. as a land of immigrants (for example, Ellis Island in New York). Clearly, in the field of historic preservation, it is in the interest of the state and of society to influence the opinions of the minorities.

Heritage organizations such as Heritage Preservation aim to bring together and keep all collectable objects for present and future generations. Their recommended programs and guidelines are intended to preserve and maintain the following: historical documents, books and archives, works of art, photographs, architectural styles and movements (in document form), monuments, anthropological works of art, historical objects and family inheritances, as well as scientific specimens, experiments or specimen copies (Heritage Preservation, Br. from 1999). Heritage organizations such as the Cherokee Indian Education Foundation, represent with accuracy cultural traditions, customs, history and language. The U.S. Heritage Foundation (1985) aimed above all at making heritage a part of the school curriculum. Historic preservation was presented thematically, both explicitly and implicitly. The relevance to a particular region, however, remained theoretical. This social-historical priority can complicate an analysis of U.S. cultural and physical heritage. It remains, however, that heritage may be interpreted similarly in both the field of historic preservation and that of genealogy.[31]

[30] At various points in his work, Alba (1990) claims that, from a purely statistical perspective, it is no longer helpful for third generation descendents to lay claim to particular regional origins, as a total of 16 great-great grandparents are involved.

[31] This is the case with the birthplace of Malcolm X. Here, the house's heritage value is as a birthplace and not because of its architectural features.

3.3.2 Heritage and Regional Significance

According to the previous definition, heritage exists in cities, libraries, museums, archives and in historical areas. The link between architecture and heritage exists in individual cases, mainly when specific architectural movements or famous architects are involved. The focus being particular cultural elements, it is difficult to detect a regional significance. Although the heritage organizations do not understand heritage in the regional-geographical sense of the word, the preservation of individual houses or structures is also a part of their function. Feilden (1997, p. 1), believes the sum of the architectural and cultural dimensions present in an historical building to be an integral part of heritage and, consequently, also of culture. If this were not so, a building would be little more that a cultureless shell. According to Feilden (1997, pp. 1), the dimensions of an historical building can be seen to possess the following attributes: architectural, historical, documentary (of a particular era), archaeological, economic and social. Nagel (1999, p. 17) proposes that research results from the field of industrial archaeology be used to enrich the concept of heritage.

The U.S. National Parks are an interesting element in this discussion of heritage. In Publication 15 of the U.S. Congress, several National Heritage Areas are designated (U.S. Department of the Interior 1996, pp. 104). "National Heritage Areas are regions in which residents, businesses and local governments have come together to conserve special landscapes and their own heritage" (U.S. Department of the Interior 1996, p. 104). In order to be designated, each National Heritage area had to show that it contained a populated area in which the history of the inhabitants was evident in its structures. The populated area and the landscape can be shown to have influenced significantly the traditions and cultural values of the people living there (ibid. p. 104). In other words, the population of the land had created a landscape that now reflected its culture. Most of the land in National Heritage Areas is privately owned. The nomination of an area as a National Heritage Area in no respect prevents the land from being privately owned, which means that preservation is difficult to enforce. The 15 National Heritage Areas have, with one exception, all been designated since the Fall of 1997.[32] The designations refer to "… a new way of conserving and enjoying the nation's natural and cultural heritage" (ibid. p. 104). Visitors to the areas can take part in a number of activities, including hiking and cycling, and even experience the restoration of historical buildings and districts. The regional significance shown here is, however, only mentioned in passing and the term should only be used when referring to specifically regional elements.

Some of the National Heritage Areas are linked to the so-called Affiliated Areas (U.S. Department of the Interior 1996, p. 99). A law that was passed in 1972

[32] Eight of the fifteen Historic Areas (ibid. pp. 106) include canals or, according to historical and economic evaluations, areas with a canal-like river system. Examples include Ohio and Erie Canal National Heritage Corridor, Delaware and Lehigh Navigation Canal, Illinois and Michigan Canal (all of which were nominated as National Historic Areas on 12th November 1996).

refers to "mixed" areas in Canada and the United States. These areas are exceptional, in that they benefit from the technical and financial support of the National Park system even though they are not state-owned (and thus do not fall under the jurisdiction of the National Park Service).

The term "heritage" cannot be understood as a single or merely regional element. The term "historic preservation", on the other hand, is based on existing, physical structures such as buildings, whereas heritage can be seen as an important explanatory dimension in the discussion of historic preservation.

3.4 Contradictory or Complementary? Historic Preservation and Urban Planning

Both the urban planning and the historic preservation movements are the result of the late 19th century, in which urban development flourished due to the industrial revolution (Hofmeister 1993). Both movements, planning and preservation, were initiated by the wealthy (Birch and Roby 1984, p. 195).

Planners strove towards reform based on rational ideas and concepts, whereas preservationists initially were motivated by romantic and patriotic ideas.[33] This "tension" between planners and preservationists is described by Silver (1991, p. 73) with concrete reference to Charleston. Official authorities worked towards what were, on occasions, diametrically opposed visions of revitalized inner-city areas. The antagonism developed when it was necessary to decide between new construction and demolition, on the one hand, and reconstruction and conservation, on the other. This caused the U.S. planning movement to be institutionalized rapidly because its aims were concrete and more economically viable. It was not until 1927 that 400 American cities brought in legislation to create an urban planning board in the form of a planning commission or a planning department. The members of the private national organizations, which were set up alongside state organizations and a large number of whom were women (American Society of Planning Officials, The American Planning and Civic Association), had their own interests as white Americans at heart (Birch and Roby 1984, p. 194). It soon became clear that the best planning projects were meaningless without the participation of the American people. In the 1940s, citizens became increasingly dissatisfied with the demolition of buildings that was taking place on a national scale. Plans were made with no attention being paid to the local culture and historical values. Professor Meyerson of the University of Pennsylvania then judged a theoretical proposal intended for urban planners as unacceptable, having been implemented with unnecessary haste (Birch and Roby 1984, p. 200). Subsequently, existing structural features of neighborhoods were used in order to plan construction effectively.[34] Combined plans were drawn up in long-term plans.

[33] The first formal construction plan concept was presented in 1909.

[34] The 1966 City of Philadelphia publication, "The Negro Housing Problem: A Program for Philadelphia", is an example of the consequences of such studies. This study defined

Table 3.2. Characteristics of historic preservation and urban development

Historic preservation	**Urban development**
Historical, out-dated	Reflects the spirit of the times, modern
Preservation	Renewal
"Slowing down" of urban structures	"Speeding up" of transformation of urban structures
Static, preservationist	Dynamic, flexible
Cultural priorities	Practically-oriented, objective
Sentimental	Favors "change"
Aesthetic/cultural purpose	Economic aims (economic viability)
Unworldly	Necessary
Application of cultural history	Variety of methods
Luxury	Defined according to political criteria

Source: based on Reul (1998, p. 83, cited according to Mörsch (1985): *Denkmalpflege und Stadtentwicklung*. In: DISP 80/81 *Stadtentwicklung*) and Buchholz (1997).

The institutionalization of historic preservation did not take place on such a dramatic scale. Its immediate aim was to "Americanize" immigrants by confronting them with historical structures.[35] This work was regionally specific and selective, and was often carried out by women (Birch and Roby 1984, p. 195). This is most likely due to the fact that historic preservation was not considered to be a "real" area of work and no specific training was required. Historic preservation is a multidisciplinary field (Feilden 1997, p. 12). No General Masterplan (GMP) existed for historic preservation programs, nor was there a developmental plan and, of course, no state budget was allocated to finance them. It was not until laws were passed in Williamsburg, in 1924, and Charleston in 1931, that historic preservation programs gained more widespread support. However, the intention here was not to found museums, but to create living, historical city districts. It was consequently necessary to gather statistical information, set up a system of historical zoning and make materials available to help with the preservation programs. Table 3.2 summarizes the main differences between historic preservation and urban planning.

Efficient planning is vital, due to the conflicts of interest that arise as a result of land use (Cullingworth 1997, p. 113). The American anti-urbanization movement, which was at its strongest in the post-war 1950s, meant that people were moving en masse to the suburbs, leaving the inner-city areas sparsely populated and allowing segregation and social divisions to become more widespread. At the heart of the urban planning projects were the demolition or rebuilding of the slums, new

neighborhood values and incorporated these into the city's planning project. This can be seen as a positive reply to Jacobs' (1993) 1961 appeal for new and better measures for historic preservation.

[35] This process would today more commonly be referred to as a process of assimilation.

construction, freeways and the renewal of the CBDs. The maintenance and preservation of old structures and buildings can hardly be said to constitute major planning concerns (Birch and Roby 1984, p. 197). Historic preservation as an intrinsic part of urban planning is a more recent phenomenon (Attoe 1992, p. 297). According to Attoe (1992), historic preservation "… remains a troublesome aspect of urban planning" (ibid. p. 297). Renovation, neighborhood renewal projects and the protection of certain areas are the aim of historic preservationists. Their methods (the inclusion of key architectural features, design plans for the surrounding neighborhood) have developed over the years into an attractive source of material for urban planners.

On the other hand, building plans and planning practice could be incorporated with ease into the concept of historic preservation. Feilden (1997, p. 2) points out the necessity of cooperation between preservation programs and urban planning in order to respect the nature of the historic structures. Not all would agree with this opinion. Many believe that historic preservation is of little importance and serves only to impede new developments and the free market economy. A number of publications thus do not mention historic preservation as an element within the field of urban planning (Chapin and Kaiser 1979).

From an historical point of view, historic preservation and urban planning can be seen to have followed completely different priorities and catered for different social classes and individuals. Moreover, they adopted individual institutional and organizational structures (Birch and Roby 1984, p. 194). In recent years, however, historic preservation and urban planning have joined forces. Thus, since 1960, planners have moved away from analyses focusing on cities and regions, and have instead concentrated more on analytical studies of the neighborhoods. At the same time, historic preservation has developed outwards from individual projects focusing on the preservation of single historical structures, and now aims to examine and support city districts and neighborhoods (see Figure 3.1).

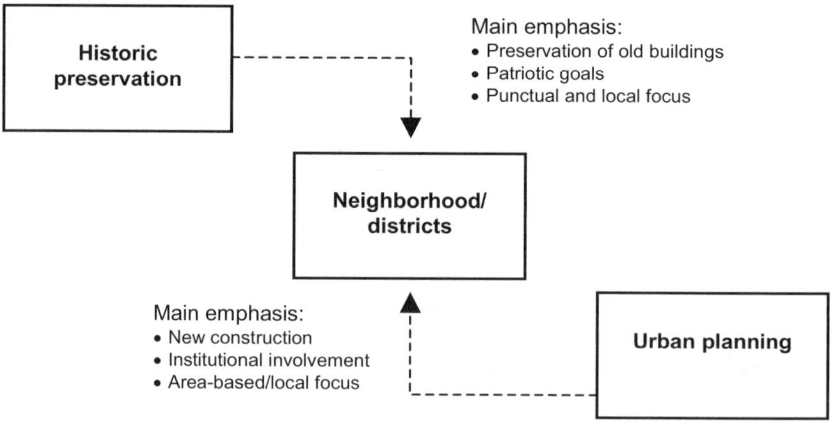

Figure 3.1. Change in priorities of historic preservation and urban planning in an historical context

Consequently, when the interests at stake are sufficiently similar, both groups work together on selected projects. Many of the historic districts were and are of particular interest to planners, both in an historical and a modern-day context. In many cases, planning projects in the historic districts could only be carried out with financial aid from planning funds (Birch and Roby 1984, p. 195). The National Historic Preservation Act of 1966 and the Tax Reform Act of 1976 did much to cement the alliance between historic preservation and urban planning.

The many building styles and social groups evident in cities can be considered from two different perspectives. Because of the presence of so many people, with varying lifestyles and cultures, an enormous "civilizing" potential exists. Individual identity can only develop once the "other" has been rejected. In an American context, this means that ethnic and social segregation can be observed. This occurs as a result of exclusion, a lack of effective planning and the social ideal of private capitalism. Urban economical studies, such as that undertaken by Park and Hoyt (see Theodorson 1982), focus on the underlying reasons and dynamics at work in urban regions. As a result of the aforementioned forms of isolation, ghettos were to develop in urban areas. In order to achieve complete renewal, in other words, a general revitalization of the entire city and not merely a few areas, it is necessary to start over by eliminating the ghettos. Total renewal and the creation of an 'open' city can only take place if difference is rejected. The main problem with this theory is that by rebuilding social structures, individual groups will not merely coinhabit an area, they will also compete with one another over the issue of access to the urban region (Korff 1991, p. 364). Exclusion, as in the case of the foundation of historic districts, is not just an emotional and psychological phenomenon; territorial control is also an implicit issue and can lead to the development of ghetto areas. As with the slums, historic districts are often urban areas, commercial districts or housing estates where ethnic and social segregation can clearly be seen. The gentrification of the former inhabitants homogenizes the existing area and simultaneously impedes further development. Today, the dividing line between historic preservation and urban planning is no longer clear (Cullingworth 1997, p. 113). A further aim of historic preservation is to unite planning techniques and the ideals of the Urban Design School (Venturi et al. 1997). Although the latter was developed in the 1960s and 1970s, Urban Design has come to the foreground in recent years and is now seen as an acceptable part of the urban planning process (see Garvin 1996).

In addition to the issues raised by the desire to create "open" cities and the problems this entails, we must also take into consideration the preservation movement that focuses on rarities. Common structures that are "surplus to requirements" are often overlooked by preservationists and urban planners alike. The preservation movement may lead to some historical periods and structures being overvalued. After all, historic preservation and historical reflection are also, to some extent, (socio-)political issues. Certain periods of history are particularly emphasized and preserved, whereas others are pushed out of the limelight. The image of the city can change according to what is shown and what is suppressed.

Philadelphia and Boston, for example, possess a wealth of old factory buildings in the inner-city areas, as well as a significant number of Victorian buildings.

Moreover, the University City of Boston contains a notable number of immigrant houses which continue to house new generations of immigrants. The calculated and theatrical manner in which Boston and Philadelphia have introduced historic preservation programs is of particular note. Contrary to the two southern cities (Charleston and Savannah), a stronger emphasis on the antebellum period is evident. Hence, in Boston, a Black Heritage Trail has existed for some years, whereas in Charleston and Savannah, there has yet been only few attempts to introduce projects (museums, trails) focusing on African Americans.

Historical structures and architectural features in both Charleston and Savannah are very different. "Master" houses can be seen alongside the cramped huts that housed slaves in inhuman living conditions. But a few of the surrounding plantations now remain. The National Trust for Historic Preservation has renovated one of the former master houses so that visitors can walk around and enter the surrounding buildings. This type of historic preservation shows an historically accurate, if "undesirable" version of regional history.[36]

Landmarks dating from the Civil War period also show the extent to which American preservation and history are selective. Such features are maintained by the National Trust. Many such areas are organized and controlled by national institutions. The continuing conservation of historic houses is evident in the materials used to ensure this. The social motives behind preservation can be identified according to the historical experiences and the specific geographical locations involved.

The preserved structures within the various cities can be examined from three historical perspectives: urban history, local history and public history. The urban processes and projects will be dealt with in more detail in the analytical section of this study, as will the transformation of local history and its systems of interpretation. Figure 3.2 should provide further insight into the different interests and planning motivations that are at work.

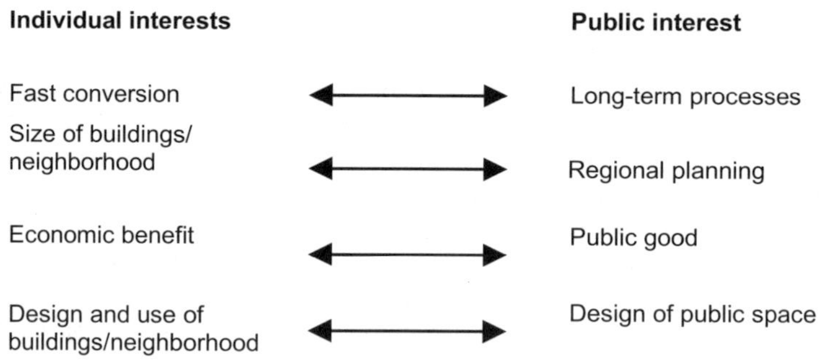

Figure 3.2. Favorable and unfavorable factors: urban planning and individual economic interests

From the existing sources, it can be seen that there are evidently several difficulties inherent in the historic preservation and planning processes. Efficient plan-

[36] Also often referred to as a "painful part of history".

ning often comes up against a pervading passivity in the United States. Moreover, it is often the case that "ethnic elements" are important regional determinants and this suggests that the cultural characteristics of ethnic groups should be taken into account by historic preservation schemes.

A further question that arises in this context is whether preservation through planning is, in fact, a viable concept. Assuming that historic preservation is a backward-looking development, we should also ask whether projects can be implemented and whether these can fulfill present needs. Planning in past eras always raises the issue of ethnic distribution, which must be interpreted and dealt with in turn by urban planners and investors. Hence, urban planning can be a political medium and this can constantly give rise to social conflict. However, it remains to be seen to what extent architects' work can, in individual cases, be considered political.

Currently, a wave of nostalgia is sweeping across the historic districts of the U.S.A., a social reaction to the renewal programs being implemented in inner-city areas, which means that more and more Americans are moving back to the cities. Thus, people are returning to the formerly neglected urban areas, bringing with them the possibility that the suburbs may once again become an integral part of the city network. Yet the historic city centers originally evolved to cater for a completely different social group, which suggests that the demands being made of the architecture and structure of the buildings are no longer the same. This makes it unlikely that society will accept an entirely authentic version of historic preservation.

The shopping malls in the suburbs today seek to "imitate" urban (infra-)structures. As a consequence, cities are now obliged to emulate the atmosphere of the malls in order to attract enough shoppers to the inner-city area. Big buyers from the suburbs are particularly targeted. It is especially important that the inner cities free themselves of the prejudices surrounding them. This privatization of public space nevertheless constitutes a segregation within the buyer group, which can be seen in the malls from, amongst others, an ethnic perspective. Unused and empty land in the cities can no longer be viewed as a symbol of failure on the part of the planners, but rather as potentially useful. Hence, it is no longer possible to plan all facets of the city. Modern needs are better satisfied by slower developmental processes (Karutz 1998, p. 24).

On the whole, the current planning situation in the United States is very subdued. The U.S. authorities have only limited political influence and, consequently, a restricted budget. Very little money is spent in total on urban planning. Americans' individual interests often take precedence over the interests of public welfare. Historic preservation should therefore be carried out in order to improve the city's image and cater for the entire population.

Although some historic districts have been renovated efficiently and well, their success is not always rated so highly. Whereas preservationists criticize the projects' transparently economic objectives, planners are often dissatisfied with the lack of attention paid to the need for progress. It is difficult to strike a compromise between these two standpoints; nevertheless, it is possible to work towards cooperation, thereby ensuring that residential regions can develop in the city, while at the same time preserving historical structures. Whereas urban planners would talk of neighborhood conservation, preservationists would describe the identical situation in terms of an efficient program of neighborhood preservation (Birch and Roby 1984, pp. 204).

3.5 Synergetic Effects: Historic Preservation and Urban Planning Today

It is the duty of the urban planner to take into account current land-use needs and their dependency on social and cultural values, and then to create relevant, detailed concepts to present to the population (Cullingworth 1997, p. 113). The inclusion of urban social elements that have developed as a result of past planning projects or settlement tendencies is today the job of planners working to preserve historical structures. If well-founded, plans which incorporate historic preservation can be superior, in that they reflect and reinforce social interests (APA 1992, p. 32). Finally, the preservationist must judge the significance of what remains and find a means to protect these fragments of the past. Existing historical buildings should be preserved for future generations, so that they can take pleasure in them and learn from them. At the same time, investors and planners must cooperate as far as possible to find a lasting function for these buildings and structures. The factors and processes that lead to a decision on the buildings' preservation have different interests at heart, while the economic considerations, scale and level of authenticity involved can vary greatly. The analytical content of this study will provide more concrete information on this subject (see also: Iv-Phl-11; Iv-Bos-1).

Historic preservation planning projects allow public institutions to work together simultaneously involving local authorities, including, above, all, the neighborhood organizations (APA 1992, p. 33). In the long term, historic preservation reinforces the U.S. urban social communities and neighborhoods. In this way, history serves to define social identity and allows location-specific identities to exist. Certain historic preservation institutions have therefore a strong motive to include minorities in the preservation projects, as mutual agreements create a greater sense of responsibility towards the neighborhoods (see also: Iv-Phl-11; Iv-Bos-1).

Historic preservation within the field of urban planning must also deal with related economic and political issues. It can also act as a means of bringing under control influential exogenous urban factors, such as globalizing tendencies, which are often taken into account in individual planning drafts. Thus, the population is included and inhabitants of the historic districts can achieve a positive sense of identity within their residential area. Cullingworth (1997, p. 114) describes this process as a reaction against the sterile International Style of more modern architecture.

A further reason to adopt historic preservation as a planning tool arises in examining the concept of renewal. A city's tourism industry and economy can clearly be improved through preservation projects. We might also mention economic and political concerns, if the deterioration of monuments and landmarks is involved. Consequently, the maintenance of less attractive monuments is often handed over by state authorities to national organizations, with the administrative responsibilities delegated upwards to higher authorities. This is generally due to a the local organizations' lack of interest in that specific part of local history, whereas national institutions can view the object in question with greater neutrality. This is often the case where disputed historical events are concerned, such as the forced demolition of slave houses in the southern U.S.A.

The patriotic reasons behind historic preservation planning projects must also be mentioned. The members of the National Society of the Daughters of the American Revolution (abbreviated as NSDAR or DAR) are a particularly good example, having emerged as pioneers of the preservation of colonial buildings and landmarks at the beginning of the 20th century.

The planning movement and the historic preservation movement in the U.S.A. are now, in many ways, run along the same institutional lines. Yet the planning movement dates back further and developed historically in different ways to the historic preservation movement. Both disciplines pay particular attention to the quality of the urban environment. This shared focus means that it is only natural that the two should work together (Birch and Roby 1984, p. 194), even if their efforts were at times at odds and, to some extent, still are.[37] Over the years, a symbiotic relationship has developed between the two movements, which is especially apparent within the historic districts.

3.5.1 Transformation in Meaning of City and Suburb: the American City

According to Goldfield (1990, pp. 26), U.S. urbanization can basically be divided into five phases: the colonial era (until the Declaration of Independence in 1776), an era of expanding national economic growth (1790–1870), the industrial city (1870–1920)[38], the suburban city (1920–1970) and the multi-centered metropolis (1970 onwards). It can be claimed that relatively little attention is paid to cities within American society. An analytical study hoping to examine the perceptions of individual neighborhood residents and their relationship to their environment must focus on smaller regional elements. For this reason, the analytical part of this study will concentrate on specific urban areas.

America's high vertical and horizontal mobility, attitude to life and mechanical resources gives rise to questions such as: Why should cities exist?, which very quickly become the dogma of U.S. urban attitudes. American society is questioning the very nature of the city as a desirable place to set up home (Lichtenberger 1986, p. 394). In the 1970s, the inner-city areas became synonymous with poverty,

[37] When discussing historic preservation-oriented urban planning processes, Ellin (1966, p. 66) on several occasions refers to planners' "anarchist convictions".

[38] The 1920 U.S. Census, for example, for the first time documents a larger number of residents than the population living exclusively within the city's SMSA.

violent crime and social alienation. The debates continue to rage even today as to whether pull factors (state subsidies, affordable suburban housing) or push factors (racial discrimination, redlining[39], ghetto rioting) are to blame for this. The unpopular idea of the "livable city" will be discussed in this work, taking as an example the historic districts concept. They represent the possibility of creating more habitable, more attractive and livelier cities as a result of historic preservation programs, and of finding new ways of improving urban microstructures, mesostructures and macrostructures.

Figure 3.3 (schematically) shows the change in meaning of city and suburb in the context of U.S. cities. Here, the historic districts play a significant role, as they alter this schematic representation by bringing a new sense of importance and central stability to urban inner-city areas.

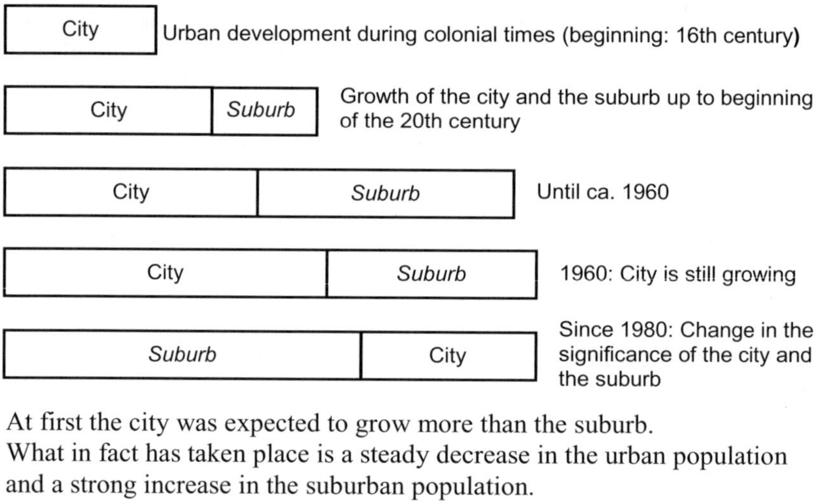

Figure 3.3. Change in meaning of city and suburb in the U.S.A.

---

[39] "Redlining" refers to the deliberate avoidance of certain areas by banks and insurance companies. Hofmeister (1999, p. 17) notes also the limited credit options offered to the residents of these areas.

3.5.2 Chronology: The History of Historic Preservation in the U.S.A.

The origins of historic preservation may be traced back as far as 1749: Peter Kahm, a Swedish botanist and researcher, tells of the preservation, due to its historic interest, of a wooden hut dating from the time of the first settlers and pioneers (Murtagh 1997, p. 207). In the early years, historic preservation work in the United States concentrated mainly on conserving individual objects of interest, such as Philadelphia's Independence Hall. These first instances of preservation were almost entirely financed by private individuals and took place mainly in the cities of Boston and Philadelphia.

Besides the preservation of individual buildings, the next milestone can be dated to about 120 years later, when the first Preservation Societies and other private organizations were founded in order to preserve and protect America's cultural history. The first organization of its kind was the Association for the Preservation of Virginia, which was founded in 1888. Before private organizations and active preservation measures were established, historic preservation was more a spontaneous and reactive process (Hodder 1999, p. 439).

In 1885, The Trustees of Scenic and Historic Places and Objects organization was founded in the state of New York. This early society developed into a national institution, thanks to more advanced means of communication and a growing interest in its aims, and was to become the American Scenic and Preservation Society (Murtagh 1997, pp. 209).

Finally, in 1906, Congress passed the Antiquities Act. This authorized the U.S. president to take measures to protect archaeological foundations and national monuments within the United States. This law represents the state's first attempt to act to preserve key landmarks and areas. Nonetheless, the law does not interfere with Americans' right to private ownership (APA 1992, pp. 3).

1910 saw the foundation of the first Historical Society, namely the Society for the Preservation of New England Antiquities. There followed a time when local governments strove to preserve historical buildings and objects, thus protecting several significant works and landmarks. It must be noted, however, that these activities had only a very rudimentary and elementary effect (Southworth and Southworth 1992, p. xxi).

Another important milestone was Charleston's historic zoning ordinance, developed in 1931. The object of this plan was to protect and preserve the historical landmarks and areas of Old Charleston and the newly-founded Historic Charleston District. We may see from this that the preservative measures were limited to a small district and were in no way part of a city-wide preservation scheme (Chase in: NTHP 1998, p. 7). In 1936, Louisiana passed a series of laws that looked to introduce architectural controls as well as protect historical buildings. Moreover, regional tax benefits were introduced in order to attract private individuals to invest in historic preservation projects (Gay in: NTHP 1998, p. 15).

A further problem arises when we look in more detail at cultural elements in the context of urban historic preservation, as various groups strive to introduce their impressions and interpretations of American history. For some years now, the former version of history, undoubtedly selective and incomplete as a result of processes of denial, and the execution of historic preservation programs have cooperated to create a new and broader understanding of the past. Minority groups (African Americans, Hispanics, Native Americans) have been able to participate and express themselves in more recent publications (Cantor 1981).

In recent years, the women's movement has come to play a significant role in the context of historic preservation (Miller 1992). Thus, some years ago, Boston set up the Boston Women's Heritage Trail alongside the official Freedom Trail.[40] Drawing on historical and feminist methods, this trail brings historically important buildings and structures to the attention of women. For example, it is demonstrated that historical structures designed by women architects have different features and decorative elements, and relate to their structural environment in different ways to the structures designed by their male colleagues (Berkeley and McQuaid 1989, Bathel 1996).

Starting around 1950 the U.S. historic preservation movement gained ground rapidly. This occurred as a result of the plans that were carried out from 1949 onwards as part of the Urban Renewal Program, which led to the demolition of many buildings. Soon, many plans were realized to renovate listed historic buildings through public funding.

Table 3.3. Restoration of listed historic buildings since 1949 in the U.S.A.

Year	Number of listed buildings
1947	Very few
1952	640
1956	1.684
1962	4.000
1979	42.000

Source: Birch and Roby 1984

[40] A further historic trail in Boston is the Black Heritage Trail, created and administered by the National Park Service.

3.6 Historic Districts as an Element of Urban Planning

The various motives behind the preservation of historic buildings will be considered further in the following section. The links between urban planning and historic preservation play a key role.

3.6.1 Special status in the United States: The historic districts

The laws passed to set up historic districts in the U.S.A. enabled the creation of a genius loci (Tunnard 1975, p. 99). Not all structures must necessarily conform to the official definition of what constitutes a building of historical or architectural significance. In the historic districts, a large number of buildings, objects and landmarks were classed together under a regional and preservationist administrative policy. Consequently, the relationship between buildings as regional geographical units, their individual characteristics and functions should not be disturbed, and their original qualities should be preserved as they were when they were originally constructed. Today, the historic districts can be viewed as by far the most active element of U.S. historic preservation (see also: Fitch 1995, pp. 41).

In 1926, the first historic district was set up by John Rockefeller in Williamsburg, Virginia. The first public historic district was created in the southern U.S. in 1931 under the name of the Historic Charleston District (later to be renamed the Old and Historic Charleston District). Others were to follow, including the Vieux Carré in New Orleans, in 1933. One of the primary objectives of the public historic districts is not to set up museums, but to preserve the natural environment; in other words, it is important that the historic districts take into account the functional and vital needs of the inhabitants (Tiesdell et al. 1996, p. 3).[41] In this way, public space is transformed into "room-like" public space within the urban area (Ellin 1996, p. 67). Hence, historic districts represented a new concept pointing to the developing tendency towards finding a compromise between existing structures and existing, older plans.

The historic districts concept was applied nationally in the U.S.A., although mainly since the 1970s (Tunnard 1975, p. 99). It was not until then that other social classes began to show an interest in the preservation of older districts and a desire to protect themselves from the consequences of the Urban Renewal Movement, which at that time was planning to demolish completely many existing buildings. The main motivating factor that pushed urban administration to intervene was the idea that historic districts could have a positive influence on the tourism industry and, consequently, the economy.

[41] The creation of the historic districts qualifies the U.S.A. as a pioneer. As early as 1956, Bornheim (1956, p. 96) notes the U.S.A.'s attempts to achieve an overview as an integral responsibility of historic preservation schemes.

It is clear, however, that the historic districts concept is not an urban one; Datel (1985) quantified and analyzed the historic districts in Philadelphia, San Francisco and Washington, DC. An interesting fact that emerged from his research was the fact that only 6% of the historic districts (of which there were eighty-five) to be found within the Statistical Metropolitan Standard Area (SMSA) of Philadelphia also fell within the city limits. Similar results were found for the other cities and these figures lead us to speculate on the activities of the inhabitants, as there can be no doubt that the older and truly endangered structures were to be found within the city limits. We must, however, look in more detail at the focus of such investigation in the course of this study.

The nomination of a particular urban area as an historic district takes place under the following circumstances: once a public committee has carried out a study on a particular area and gathered statistical information regarding its structural content, this is recorded as detailed, written documentation (Tunnard 1975, p. 101). In many cases, the local neighborhood organizations are behind such initiatives.[42] Next, a commission is set up through the local government, which, in most cities, is called the Board of Architectural Review (BAR). This committee is responsible for deciding on overall historical design concepts, smaller integral features (such as the color and height of surrounding fencing) and the materials and color combinations to be used. These binding decisions do not take into account structural form, the height of the buildings in relation to their surroundings, interior features (staircases, cellars, attic extensions etc.), features specific to the building in question or elements which are invisible at street level. This is in the interest of protecting individual rights in the United States. Any supplementary construction work undertaken on existing structures which are visible from a public street or right of way must first be approved by a commission called by the city council. This applies where demolition or partial demolition is involved, as well as in the case of visible poster advertising within the designated historic districts.[43] This is also true in instances where colors and logos specific to certain companies are involved, a marketing policy which allows passers-by to recognize the firm at a glance, as in the case of food or restaurant chains. This conflict of interests between "swift recognition" and the "individual character of the neighborhood" with reference to company logos is at the heart of Millard's research (1991). A large number of food and restaurant chains must comply with the historic districts' strict regulations and omit company-specific colors.[44]

The first moves towards converting a neighborhood into an historic district is often made by the residents themselves. The process by which it is decided how far the ideal of historic preservation is followed, in historical and architectural terms, varies from state to state. Massachusetts is often cited in literary sources as an exemplary U.S. state which deals effectively with this issue (Tunnard 1975,

[42] This was the case in the examples of Society Hill (Philadelphia) and Beacon Hill (Boston).

[43] See Figure 4.7.

[44] It must furthermore be noted that some companies use these historical factors in their advertising campaigns within the neighborhoods.

p. 101 and Maxmann 1994, pp. 4). The analytical element of this study will look further at specific structural examples.

3.6.2 Globalization and Historic Preservation

Following the process of globalization, we can today look forward to the development of universal time, economic, cultural and planning systems. Nevertheless, every country has its own culture, its individual traditions and a unique history, and national, regional and local identity cannot be altered so easily. In the U.S.A., this cultural and traditional pluralism is evident in regions where specific buildings and groups of settlers are to be found. The U.S.A. is particularly interesting from a preservationist perspective because its features and structures have developed over a relatively short period of history, which leads to continuous internal and, in this global era, external processes of renewal.

Goods and ideas are constantly being exchanged on a massive scale, whilst many see social changes (Clark 1996, p. 3) as a clear indication of globalization. In his monograph, Clark (1996) points out the contradictions inherent in modern life, where globalization seeks to unite different cultures and individualist tendencies work against such a union. Korff (1991, p. 357) believes that global and local processes are two facets of a single, greater process. Thus, he speaks of international integration together with the processes of diversification and localization.

For the urban area, globalization represents change, so that production must now take international markets and competition into consideration. Some companies have developed into global players, whereas others have formed strategic alliances through merger deals. Today, there is competition between cities to gain access to capital that has been virtually immobilized. As well as the widespread exchange of technology and knowledge, financial markets have been internationalized to a great extent (Sassen 1991 and 1994). The globalization and internationalization of many procedures have redefined what we see as our own and what we perceive as foreign. Global integration also means that less attention is paid to the individual.

When mentioning globalization, it is important to conceptualize the definition of the term. In the early 1990s, it did not appear in English or German dictionaries and has only in recent years been adopted widely by specialists. Experts have now begun to define and explain the globalization phenomenon. In the field of regional research, many examples can be found of new terminology for which no clear definitions exist (Strubelt 1999, p. 67). This is the case with the term "sustainability". This, originally a forestry term, is now used widely within the fields of natural science and sociology. Globalization a term developed as a result of economic analyses of international economic processes and describes the worldwide integration of economic relations and systems of exchange created by increasingly efficient means of communication on an international level. Yet, even today, it remains unclear, due chiefly to a total lack of relevant research, whether these developments are the result of economic factors alone, or whether geographical factors are also involved, which are reflected in the local or regional zeitgeist by certain architectural or cultural features. If this is so, it indicates a widespread interdependence between different geographical areas, for example, the development of an alternative workforce in so-called Third World countries (Friedrichs 1997). Hence, globalization refers to the current development in international relations. This is not a new phenomenon, but represents rather the attempt to "...reflect a development through description and the definition of norms, without in any way allowing us to develop a vision of the future" (Strubelt 1999, p. 70).

Some may suggest that globalization is not necessarily to be seen as a westernizing or Americanizing process; instead, cities themselves are striving towards more popular and more local forms of identity through diversity, variety and multiplicity. As a direct consequence of the growing similarities and common characteristics between different cultures due to the new media and the accompanying over-regionalization of individual architectural features and landmarks, historic preservation within the field of urban planning is experiencing a renaissance. The priority now is to prevent a region from being 'imitated' by preserving and maintaining its unique features. This means that globalization can also be beneficial for a region when relevant and well-organized urban planning programs follow.

In the context of globalization, historic preservation can been seen to play a further role, as a means of protecting the native country. The preservation of historic buildings often reflects past eras where globalization did not yet exist (Hodder 1999, p. 439). The general question as to whether societies will return to a regional orientation or embrace international universality remains as yet unanswered. The concept of so-called global cities (Sassen 1991 and 1994, Friedrichs 1997) is thus merely a means of explaining particular urban phenomena and fails to include current trends that lean towards urban political historic preservation, heritage and culture, a popular idea that, in fact, represents little more than the tip of the iceberg.

In this study, it is interesting to note that globalization forces competing city regions to merge, whereas the latter nevertheless succeed in preserving their individual identity through historic preservation. The global phenomenon cannot be said to affect all cities in the same way (Korff 1991, p. 358). Nor is the whole of society affected, so that historic preservation is also partly a (political) means based on economic factors and the promotion of corporate identity, promoting local identity and respecting preferences that are specific to an area. Variety is what brings a city to life, whether or not it is particularly affected by globalization, different communities or location, and this variety is what gives it its main characteristics (Korff 1991, p. 157). Today, a modern city should ideally be organized on a local basis and simultaneously be shaped by globalization.

Historic preservation makes it possible to combat the downside of globalization. Regional associations should cooperate, as is the case with tourism. Thus, a region can cater to tourists who are looking to stay for several days rather than make a day trip to the area. At the same time, this allows for a greater number of visitors, as the travel time to the area can be significantly longer. Well-organized strategies should be planned out to avoid potential traffic problems caused by the globalization trend.

The depopulation that has taken place in most American cities means that a great deal of money has been taken away from the immediate city region. The following section will show the extent to which historic preservation programs also suffered as a result. During the Reagan presidency, there were even years where no support whatsoever was won in budget debates (Birch and Roby 1984, p. 204). It was also because of such financial cuts and the preservationists' poor image that urban developers consciously excluded historic preservation from their projects.

3.6.3 Private Capitalism, Individualist Tendencies and Historic Preservation

In German-speaking countries, it is above all this Lichtenberger (1986, p. 392, p. 394 and 1998, pp. 154) who has dealt with this terminology of politics and administration. She stresses the social economic tendencies that underpin urban research processes, and refers particularly to the phenomenon of private capitalism (1986, p. 392). The city's physical structure becomes instead a static region defined by data. An immediate concern with financial gain becomes the focus of this particular style of capitalism, whereby urban development is determined by the economic competition between active employment markets (Lichtenberger 1986, p. 392).

When the first settlers arrived and founded the first U.S. cities, these settlers from different religious (mainly Protestant) and cultural backgrounds had their own definitions of politics and society. Baltzell (1980) deals with this in his monograph, taking as an example "Puritan Boston and Quaker Philadelphia". Putting all other differences aside, the settlers agreed to work towards a secular form of private capitalism. The latter's roots are to be found at the very beginning of the European settlement in North America and thus a long tradition of private capitalism exists in the United States (Burkart 1993, p. 172).

The financial and regional authorities' power to make decisions lay in the hands of individual members of society who, in psychological terms, strove towards the happiness and satisfaction of individual freedom, and wealth. U.S. liberalism reinforced the concept of freedom of choice in a political and administrative sense. From a social perspective, private capitalism meant that individual members of society saw the well-being of the family[45] as their most vital concern. From a political standpoint, private capitalism was directed only to the extent of enabling peaceful and untroubled exchanges to be carried out between traders within the community. On the one hand, earnings, systems of employment and net social production output depended directly on the thousands of individual enterprises that flourished in the U.S. states, and not on a general, centralized – i.e. state – organization. On the other hand, private capitalism also influenced the physical features and outward appearance of cities, with their factories, offices and residential buildings: empty lots were mostly allocated to real estate agents and

[45] Later, this would be extended to include the neighborhoods (Weiler 1974, pp. 10).

land speculators who were aiming to make a profit (Warner 1991, pp. 85 and Lichtenberger 1998, p. 218, pp. 287). Private capitalism was to set up a conflict of interests, as far as historic preservation was concerned, between the public sector and private investors. The main interests at work have been summarized in the following table:

Table 3.4. Main state and private interests affecting historic preservation

State	Private Investors
History	Aesthetics
Significance in general	History
Promotion of culture	Significance on a personal level
Aestheric concerns are not necessarily a priority	Individuality
Selection criteria ranked in order of significance	Creativity
Cultural use (museums etc.)	Profit

These interests are particularly marked in the field of historic preservation. The clearest conflicts of interests and contradictory objectives are shown in Table 3.5.

Table 3.5. Conflicts of interest and contradictory aims: public sector vs. private sector

Factors	Private Sector	Public Sector
Main Motivation	Return on capital, profitability	Political
Who makes decisions	Centralized	Varies, socially oriented
Goal	Clearly defined, short-term	Extremely wide, often long-term
Selection criteria for alternative options	Specific, economic parameters	Ambiguous motives (e.g. economic)
Authority over lesser urban measures	Unclear/changeable limits and measures	Local authorities
Temporal focus	Short-term	Long-term

Source: based on Mercer (1991, p. 26)

In order to ensure that all factors are taken into account, it is necessary to mention here the Transfer of Development Rights (TDR) system (Costonis 1974 and Hofmeister 1977, p. 148). This allocates particular rights to cities. The TDR system is based on the idea that the land on which an historic building stands does not use the space on which it is built as efficiently as it could (i.e. it does not take advantage of maximum height allowances), thereby significantly reducing its profitability. On payment of a fee, the owner of the next plot of land can benefit from a 20% increase in constructional zoning limits. This fixed fee is then placed in a fund and paid to the owners of the historic building on the condition that they do not demolish it. This system takes into account the fact that the majority of historic buildings are privately-owned. It must be noted that the TDR system was only successful in urban areas such as Manhattan, New York or the CBD in Chicago, where historical features are located in areas of particular interest to investors (Costonis 1974). The demand for land in such areas means that fees are readily paid and can be used to subsidize historic preservation.

In the United States, both areas of production and employment are capital-oriented, and this has been the downfall of many industries. Great horizontal and vertical mobility led to widespread segregation. This is reflected in social, demographic and ethnic criteria, as well as the various lifestyles that can be identified in the U.S.A. (Lichtenberger 1998, p. 154). These parameters are clearly to be seen within the historic districts. In the past, the middle classes, who generally look favorably on private capitalism, prevented budgetary measures from being taken to finance restoration work or social schemes in socially disadvantaged areas (the so-called slums[46]) (Lichtenberger 1998, pp. 127). It was rare that the government intervened to support the dying manufacturing industries of the Manufacturing Belt (in Philadelphia and Boston, for example). The four cities of Philadelphia, Boston, Charleston and Savannah, examined in detail in the analytical part of this study, are also subject to this tradition of private capitalism. In 1970, Warner (1991) published a monograph, "The Private City", in which he uses Philadelphia as the basis for an analysis of the social and economic structures of private capitalism. The parameters he describes can equally well be applied to Boston, Charleston and Savannah.

The exact social cost of private capitalism is unknown, but undoubtedly extremely high. As early as 1950, Kapp published the results of his investigation into this matter. Unemployment and social segregation (in addition to the already existing ethnic segregation) became widespread and the disparities more marked. Historic preservation is a way to ensure that investors return to the neighborhoods and the cities, so that residents can once again identify themselves with the area and the city. The individual districts' crime rates, unemployment figures and residents' level of education can clearly be identified, and citizens are more inclined to cut themselves off from such social groups than to aim for a form of integration that could solve the latter's problems. This ideal can be achieved, however, through powerful neighborhood organizations.

Burkart (1993, p. 173) argues that this individualist tendency is not a universal trend, but a privilege. Although Burkhart's research concentrates on parents in the U.S., his work can be used to identify universal social patterns and regulations. Freedom of choice can only exist if a variety of options are available which allow multi-faceted forms of existence to develop. Limited education and restricted financial means dramatically reduce the options available. Hence, the maxims of private capitalism act mainly to benefit and support the middle and upper classes. These are accepted by the lower classes who subsequently imitate and support the

[46] A "slum" refers to an area in which at least 50% of the apartments in each apartment building fail to meet with agreed standards, i.e. do not contain a kitchen or bathroom with a supply of running water (Hofmeister 1999, p. 171).

ideals of private capitalism.[47] This social tendency is not weakened by the social classes in question, as they believe that social benefit would have a leveling-out effect on individualizing processes.

This question of individualizing tendencies is interesting, in that it supports the integration of historic preservation, which also reflects the ideals of general welfare. Prior to 1980, house owners did not have the right to contest entries listed in the National Register for Historic Places (NR) because this list (based on Cullingworth 1997, p. 117):

1. is organized according to subjective methods of evaluation,
2. does not allow the owners of listed places to use their property as they choose,
3. provides no comparative argument as to what does and what does not constitute an historic object

As a result of political pressure, owners now have the right to contest entries in this list of places. A compromise was found and included in the NR regulations in the 1980s, namely the concept of "eligibility for inclusion". This means that even if a building qualifies for listed status in the NR, it need not necessarily be listed. Normally, such buildings are treated much as buildings listed in the National Register (NTHP 1997, p. 53).

It has become clear in the course of this chapter that cities' systems cannot be analyzed independently of their political systems. Moreover, the argument can be turned around to show that regional elements play a key role in political systems.[48]

3.6.4 Urban Renewal and Historic Preservation

Edmund Bacon (1976), a member of Philadelphia's urban planning bureau from 1949 to 1970), believes that the economic well-being of an entire region is directly dependent on the financial situation of the inner-city districts (see also Cybriwsky 1986, p. 27). It is Bacon's opinion that, because the economy in these districts lying at the heart of the region (CBD) is so strong, investors and citizens throughout the United States have become aware of such areas and are ready to invest in them once again. Other authors, however, believe that the prosperity of inner-city areas does not concern surrounding districts and regions. Each individual neighborhood must create an economically favorable climate and attract investors. A single, all-embracing image that can apply to other districts or regions does not yet exist (Levine 1987, p. 118). Thus, the renewal of inner-city areas is realized through private financial support and, in the context of gentrification, is not necessarily aimed at benefiting socially disadvantaged residents (Cybriwsky et al. 1986, pp. 117).

---

[47] As, for example, the opinion that "life is what you make it" (Burkart 1993, p. 172). This proverb seems to find greater resonance in the U.S.A. than in European welfare state nations.

[48] In this context Hofmeister (1982, p. 487) proposed a series of culturally genetic urban types.

The key terms "revitalization" and "urban renewal" from the 1960s today refer to little other than the economic renovation of urban regions. Today, the revitalization process is interpreted much more widely and now aims to promote cultural aspects and take into account issues of historic preservation. Critics rightly question whether we accord too great an importance to the past and whether this obsession with historic preservation is ultimately meaningful. Such identity fixations often serve only to promote the city's general outward appearance and create a certain image of the city, whereas little attempt is made to deal with social aspects. Nevertheless, the process of rehistoricization can have a positive effect. Historic preservation is, after all, a means of making culture open and accessible to all. As well as residential buildings of interest, urban planning in the U.S. must take into account and make provisions for a vast number of commercial constructions within the field of industrial archaeology.

3.6.5 The Process of Gentrification and Historic Preservation

The term "gentrification" was first coined in 1963 by Ruth Glass. The term was to be applied with increasing frequency in discussions on the transformation of inner-city residential areas. Gentrification refers to the process by which the nobility moves back to the city from the country. Franzmann (in: Friedrichs and Kecskes 1996, p. 230) defines gentrification thus: ".. the entrance of a group of citizens perceived as socially superior and the subsequent revaluation of residential housing and infrastructures" (based on: Dangschat and Blasius 1990). Particularly in the United States, the term has become a hotly disputed technical term, as the definition derives from European class systems that never existed in a comparable form in the U.S.A.[49] In German and British contexts, however, terms such as revaluation, (urban) reinvasion, (inner-city) revitalization, neighborhood renewal, neighborhood reinvestment and middle-class resettlement occur (Dangschat and Blasius 1990, p. 11). The term can be applied to U.S. gentrification processes in relation to qualitative population exchanges, the changes in living conditions in the wake of modernization, the change in emphasis from rented to privately-owned apartments, reinvestment in neighborhoods and infrastructure changes within the affected districts.

Economic and demographic factors, residents' lifestyles, the housing market and the districts' political planning projects represent the parameters of the process of gentrification and the push and pull factors involved. In his study, Smith (1979, pp. 538) uses the example of Society Hill in Philadelphia to consider in more depth the phenomenon of gentrification.[50] He concludes that the successful

[49] Friedrichs and Kecskes (1996, p. 7) point out that the phenomenon was first discussed in the U.S.A.

[50] It is important here to note that we are dealing with a process and not a status quo (Friedrichs and Kecskes 1996, p. 7). Alisch and Dangschat (in: Friedrichs and Kecskes 1996, p. 95) develop this difference further in *Gentrification als Situation und Gentrification als Prozeß* ("Gentrification as a situation and gentrification as a process").

revitalization of inner-city areas within the neighborhoods depends to a large extent on the available capital and not on the people themselves. "Gentrification is a back to the city movement, all right, but of capital rather than people" (ibid. pp. 547).

"New jobs" that have developed as a result of high technology (e.g. in the fields of finance, insurance, legal advice, real estate, advertising and information processing) point clearly to a push factor in the revaluation process. These mean that new heights have been reached in the salary hierarchy and such jobs are mainly occupied by young, unmarried and highly-trained workers who are rediscovering residential areas around the city center.

The long-term displacement of the established residential classes as a result of gentrification requires that newcomers and existing residents live together in the short term as a heterogeneous whole. On the one hand, we find old-established residents and on the other, the gentrifiers (Dangschat and Blasius 1990, p. 23). All play a role in the process of gentrification. This heterogeneity, as well as the gentrifiers' own ideal of neighborhood homogeneity, cause cultural, social and economic problems. The greatest problems inherent in the gentrification process are the result of socio-economic polarities and gentrification's process of displacement, problems which should be avoided or, at the very least, whose impact should be reduced (Marcuse 1986, p. 154 and p. 173).

Whether gentrification represents "a" problem or "the" problem is a question that can only ever be answered in relation to individual cases (Dangschat and Blasius 1990, p. 28).[51] Weiler (1980) and Cybriwsky (1986) provide much material for thought in their theoretical works on the gentrification process in Philadelphia, which they illustrate with many hypotheses and examples to show that not all superficially positive outcomes prove that revaluation and improvements have taken place. The results must be considered and evaluated on a wider scale.

Gentrification is a process that, for a variety of reasons, is closely linked to historic preservation. On the one hand, the historic districts are often to be found in areas that were in need of revitalization. Moreover, this revitalization was often financed with money intended for programs of historic preservation. On the other hand, gentrifiers often possessed the necessary financial means and the relevant training to carry out historic preservation projects and at the same time profit from available tax benefits.

Public authorities must be careful to ensure that rents and prices of goods do not increase too dramatically within the historic districts, so as to avoid criticism of "insidious gentrification". Datel (1985, p. 141) sees the transformation of historic preservation into a "real-estate gimmick" as an obviously inspired means of introducing the gentrification process.

[51] Moreover, Friedrichs (1996) refers to the generally "chaotic concept" of gentrification (in: Friedrichs and Kecskes 1996, p. 13).

3.6.6 Conversion and Historic Preservation: New Life in Old Buildings

The conversion of buildings or complexes is a significant element in urban planning concepts within the inner-city areas of the U.S.A., precisely because disused and run-down buildings and industrial complexes present a particular problem in these areas. Moreover, the needs of the city's residents have changed over the years, thereby making the conversion of historic housing or offices necessary. Even within the historic districts themselves, controversial discussions are taking place as a result of zoning ordinances and conversion plans.[52]

Within the specific U.S. urban planning climate, this necessary change in urban needs is addressed by taking into account political, economic, social and architectural factors. In the CBDs, many disused historic buildings dating from the suburbanization eras (in particular the 1920s and the 1960s) exist and represent ideal resources for various conversion projects. Today, production plants, factories, warehouses, port installations and old theatres often serve no functional purpose; nonetheless, these empty, disused buildings are still worthy of preservation and are a visible part of the urban landscape. Often, the only viable renovation option in economic terms is to convert these buildings. Simple revitalization programs are usually unfeasible from an economic and urban planning perspective, and such planning programs are unsuccessful. The potential benefits of and problematic factors inherent in conversion plans can be identified through research into the historic districts. The following analysis differentiates between four distinct forms of conversion: residential, commercial, work-related and leisure conversions (Kupinski 1999, p. 139).

The conversion of buildings makes it necessary to adapt old historic structures. This gulf between past and present is used by many architects to create attractive architectural contrasts. Architects must first of all ask themselves the following questions: how do we define "preservation" and to what extent can conversion work be carried out? How "creative" can and may conversion work be?

The socio-historic values by which a building is judged can be described as follows (based on Feilden 1997, p. 6):

1. Emotional value (identity, continuity, spiritual or symbolic significance)
2. Cultural value (documented, historical, archaeological, aesthetic, architectural)
3. Functional value (functional, economic, social, political)

The conversion of old factories, office buildings or residential blocks serves not only to bring about the revitalization of deserted urban economic nuclei; it simultaneously enables the renovation of historic and protected buildings. Converted structures also allow people to identify with their home city and its historic buildings. In this context, Smith (in: AIA 1997, pp. 234) even goes so far as to identify national economic factors that lead to a renewed investment in converted buildings. Infrastructural analyses and ecological aspects (structural burdens, existing

[52] The American-English terms "conversion", "historic re-use" and "adaption" indicate the wide range of interpretative possibilities of the term "conversion" (see AIA 1997, section on "Adaptive Re-use", pp. 255).

structures) must also be taken into account when considering conversion work. In some cases, it is unclear who owns and who has a right to own certain property and this represents an evaluation and economic risk for private investors and landowners (Kupinski 1999, p. 140). Few U.S. states have taken measures to eliminate this risk.

Conversions can help towards developing a city's image. Historical structures will not be "chemically" preserved as they were originally intended, and, unlike renovation programs, this prevents them from becoming museum pieces. A former industrial city can, for example, transform and improve its image through well-organized conversion projects. Conversions renew a building's locational quality; in other words, old locational factors change when necessary and, in this way, can become either favorable or unfavorable factors. New users choose to move to the area and take into account the neighborhood's urban and historically specific environment. Currently, conversion projects in Charleston are working to transform office buildings into residential blocks and gentrification is expected to be the result (Iv-Cha-2).

When carrying out conversion work, it is vital that public and private planning offices aim to create useful, long lasting and long-term space which will continue to reflect the city's history and its individual features. An "untouched" historical structure in need of conversion in no way represents a defeat for the politics of urban development (Karutz 1998, p. 24). On the contrary, it is quite feasible for historical buildings to become profitable with the passage of time (Black in: AIA 1997, p. 255). The conversion of buildings into cultural and other public socio-cultural centers (museums and community centers) is only possible if this is in the interests of the neighborhood or the city itself (Warner in: AIA 1997, p. 237).

Conversions allow a contemporary sense of identification with the city and its converted historic buildings which come to be regarded as symbolic.[53] Planning authorities should aim to find a long lasting and long-term use for these buildings which will reflect the city's history and individual traits. In individual cases, this can be critically assessed in a number of ways, which are listed as follows (based on: Kupinski 1999, pp. 142):

1. Location, infrastructure of the building
2. Mixed use, neighborhood (contributes to the urbanization process)
3. Attractive architecture
4. Economic aspects for property owners
5. Attractive ground plans (interior architecture)
6. Flexibility of use (change from living space to work space should be possible)
7. Preservation concerns should be respected when dealing with the structure

The following chart should bring together findings and the aspects that play a role in the research processes that precede the re-use of buildings and complexes in urban regions of the United States (Figure 3.4).

---

[53] A discussion on the theory of "symbolic regional identity" (*symbolische Ortsbezogenheit*) can be found in Treinen and Rosow (1974, both authors in: Atteslander and Hamm 1974).

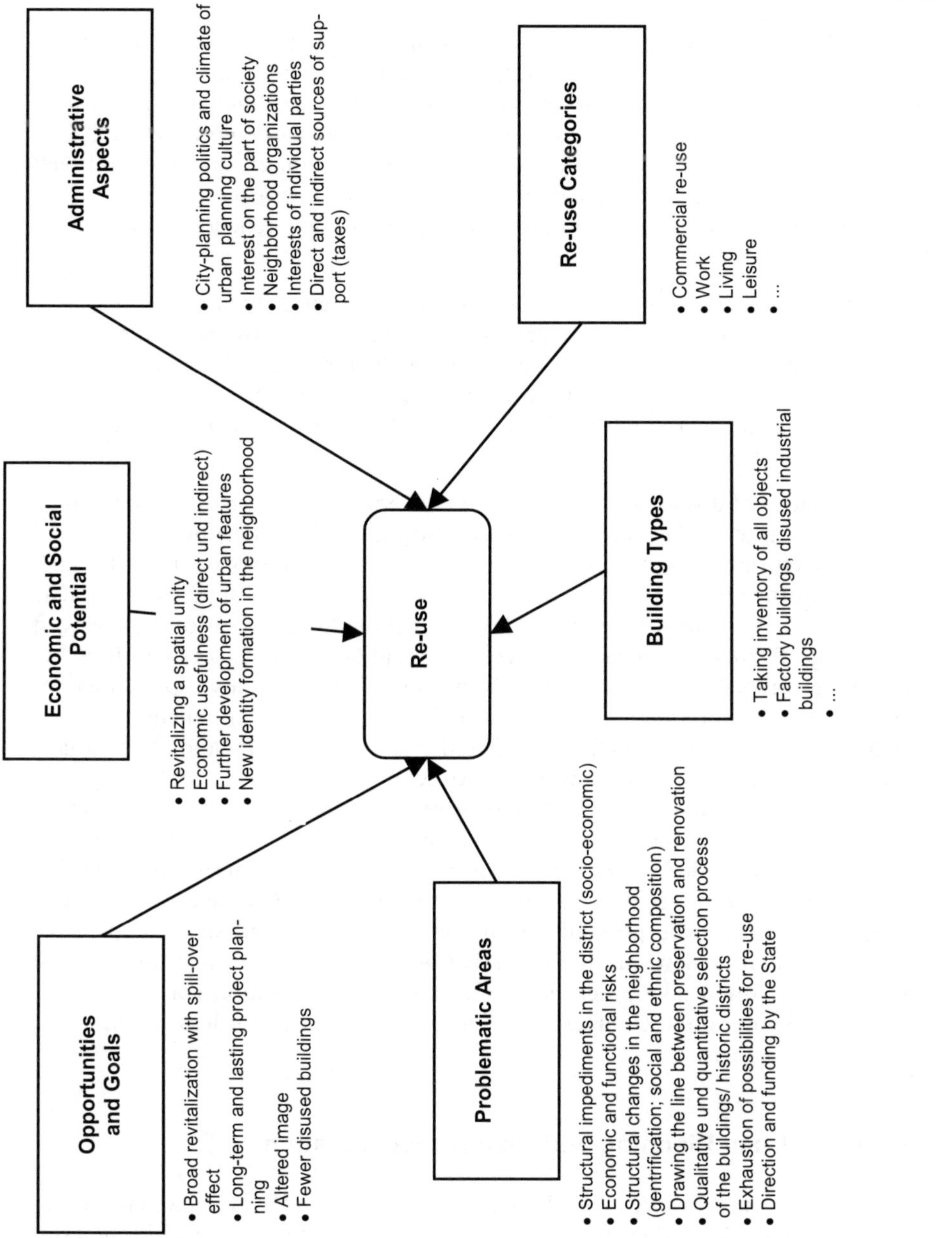

Figure 3.4. Important aspects of research processes preceding the re-use of buildings and complexes in urban U.S. regions

Sacred buildings receive special status within the field of structural conversions. On the one hand, the possibilities for converting the inner space are restricted, while on the other, such conversions might harm traditional forms of worship. Demolition is complicated for the latter reason and is often impossible due to the building's particular and individual architectural features, and its traditional cultural and social significance within the city (Iv-Phl-11, Iv-Bos-1).
A conversion or re-use leading to mixed use of a building depends on (based on: Kupinski 1999, p. 138):

1. its specific location and immediate relation to its neighborhood,
2. specific locational assessments, as a result of which it is possible that old locational requirements should be compared with a newer set of requirements (standard locational evaluation processes are not sufficient),
3. recognizing and avoiding conflicts that may arise because of the building's planned function

In the course of discussions on proposed conversions, the question of individual buildings' location (infrastructure, ecological aspects, micro-location) plays a vital role. Old factories are not necessarily to be found in locations of interest, while convertible office buildings are often located on streets with a high level of traffic and noise. Conversion renews a building's locational quality, in other words, it may change factors that formerly affected a location. Such factors can thus become either favorable or unfavorable. New users choose to move to an area because of the neighborhood's urban and historically specific environment.

Contrary to Europe, the U.S. has suffered less since its foundation from a lack of space. For this reason, urban historical structures were rarely overdeveloped or overplanned. Some dilapidated buildings were occasionally rebuilt in a few inner-city locations. Hence, the continuous development of cities is synonymous with the expansion of built-up areas. This has favorable consequences, meaning that many historic buildings in the U.S. are still standing today. On the other hand, this significantly reduces the possibilities of conversion programs, as new investors can easily obtain land within the city's surrounding green belt and questions of economic gain often take precedence over historic preservation concerns in the minds of U.S. investors. Compared to Europe, therefore, much less capital is invested in urban areas.

3.6.7 New Urban Planning and Historic Preservation

The recent tendency towards New Urbanism, identified and examined in particular detail by Smith (1996), does not merely take into account the city and its suburbs, but also looks at the ways in which the population perceives the idea of "community" and how it leaves its mark on a particular region. The multiple facets of public spaces can be seen to affect the region in many different ways.[54] New Urban

[54] Smith: "New Globalism-New Urbanism", lecture at the University of Pennsylvania in Philadelphia on October 29th, 1998.

Planning's urban planning phase describes and takes priority over the current urban politics of renewal. The considerations and problematics involved can be traced back to the ideas of New Urbanism.

Katz (1994) justifies New Urbanism as a new, lasting form of urban growth based on innovative concepts which take into account the population's new needs. Urban structures which are already in existence should, however, not be transformed completely; instead, existing structures, such as the system of streets, should be incorporated. Katz sees appropriate standards of living, infrastructures and the furthering of a sense of community as vitally important considerations.[55]

Venturi et al. (1997) painted a clear picture of American architecture's general post-modern ideas. Basing their analysis on Las Vegas, they show how and why architectural development has grown so uncontrollably over a wastefully large area of land. At the same time, they demonstrate why planners cannot simply move on and away from this "American style". Standard ground plans must not necessarily stand in the way of standard demolition plans which, to the European eye, may appear on occasions too calculating. All authors from the New Urbanism movement agree on one thing: the American Dream must be redefined (Hayden 1986). Moreover, they refer to the vision proposed by Lynch (1997), who himself was a researcher into urban image and supported the ideas of Urban Design. These advocates of New Urbanism put forward a further series of fundamental questions:

Under these circumstances, how can a new policy of growth and development be achieved in the long term? To what extent must cities seek a compromise between physical features and individual history?

Kegler (1998) points out that the Congress for New Urbanism (CNU)[56] proposes to combat this urban depopulation in a number of ways, the protection of cultural wealth being one of the greatest challenges with which humanity is faced (Kegler 1998, p. 335). Thus, the New Urban Planning charter was passed in Charleston in 1996 by two hundred participants, in which Congress referred to "the lack of investment in city centers, a general movement of settlers away from these areas, growing regional divisions based on race and earnings".[57] The demands put forth by the CNU find their basis, amongst other things, in locally-oriented architectural features and urban structures. The historical overview need not take into account local considerations. A heterogeneous neighborhood is the preferred goal (ibid. pp. 336). It is also important that the preservation and renovation of historical buildings, city districts and urban landscapes ensure the continuity and development of urban society (ibid. p. 342). In this way, it is clear that a positive fundamental philosophy has emerged that takes into account the ideas of the historic preservation movement. Kegler (1998) furthermore suggests that New Urbanism is a form of inner-city neo-commercialization (ibid. p. 345). We can

---

[55] To find out why this criterion is rarely taken into account in most U.S. cities, we can look for further information in the works by Vergara (1995) and Garvin (1996) or, for a methodological and sociological approach, by Wilson (1990).

[56] This conference has taken place on an annual basis since 1993.

[57] The complete German translation of the charter can be found in: Kegler 1998, pp. 336.

agree with this objection only up to a certain point: it should only be taken into account if the charter's provisions are disregarded.

Moving on from the results that have emerged from this thematic discussion of 'urban historic preservation', the practical focus of Chapter 4 will concentrate on those involved in historic preservation programs in the U.S.A. Priority must be given to the question as to who exactly influences the political-administrative, economic and socio-cultural structures in the historic districts within the four cities examined in this study.

4 Who's Who: Support Strategies, Interests and Influence

Different participants determine historic preservation in the U.S.A. In the following, they are divided in organizational and theoretical terms on three levels: national (federal), state and regional (local) individual states levels. The different institutions are then further divided into national, public and private institutions (Figure 4.1).

To some extent, the many different institutions operate with very different concepts and aims, which are identified and discussed in the following chapter. This study focuses on the investigation of this range of motivations and concepts, and identifies opportunities and potential within the historic preservation effort. It is, of course, not merely statistically measurable and detectable parameters, such as economic profit, rising tourist numbers or the number of successfully historically preserved houses, that play a significant role. Possible social modifications, as well as guiding and manipulative elements, must also be taken into account.

How these theoretical concepts are realized and in what respect historic preservation in the U.S.A. is directed by the participants both economically and socially is to be clarified in the empirical section focusing on the four historic districts in Philadelphia, Boston, Charleston and Savannah. The summary of this organizational-theoretical section is divided into three categories (political-administrative, economic and socio-cultural), which form the evaluation levels for the following empirical section.

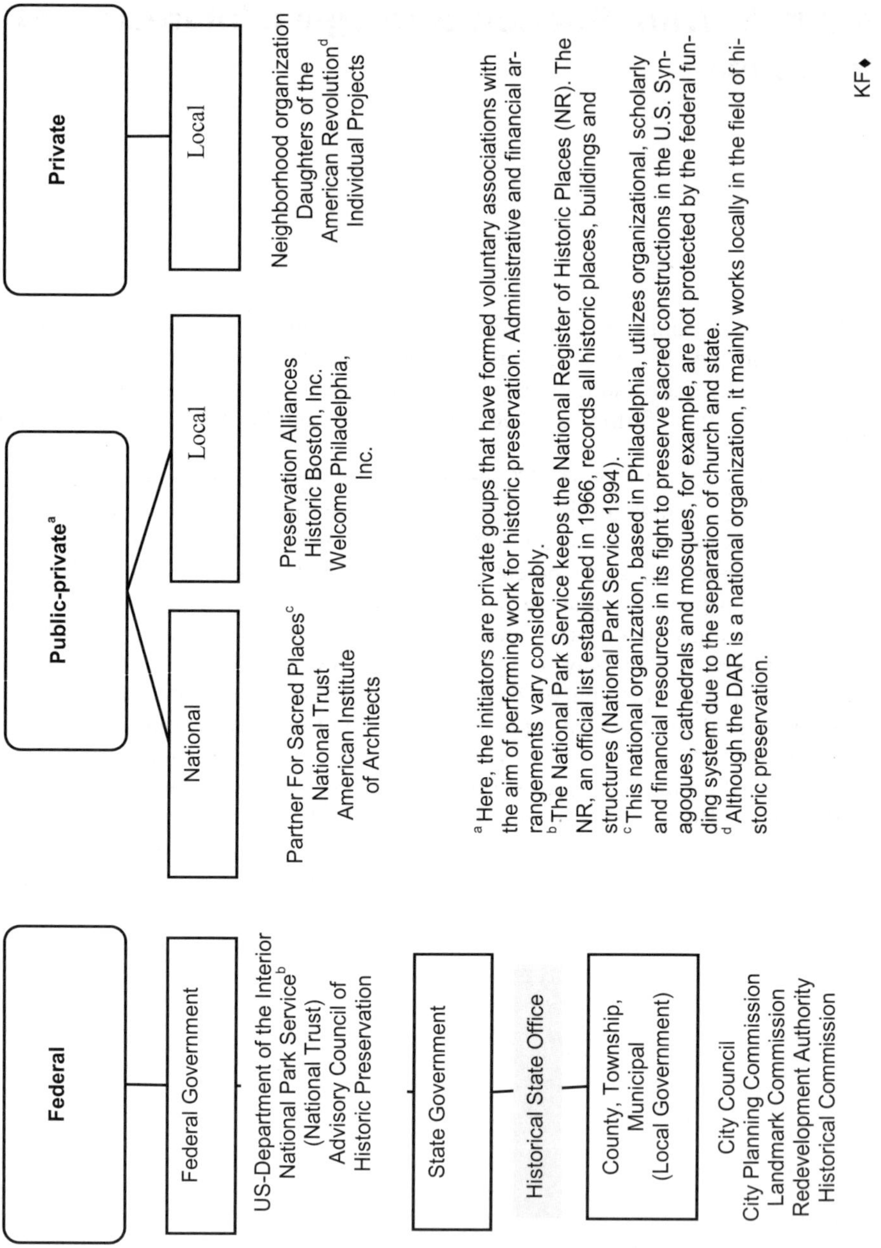

Figure 4.1. Synoptical representation of the participants in the research field of historic preservation in the U.S.A.

4.1 International Level: Global Strategy

Among the many institutions which are involved in architectural and cultural preservation as well as historic preservation in American cities, internationally active historic preservation organizations tend to take a back seat and exert relatively little influence. Nevertheless, the main organizations should be mentioned here in order to ensure the completeness of this representation. It is thus of interest to examine their global methods and targets, as well as the methods and strategies of approach. Worldwide organizations have as their goal the protection of internationally significant features or, in general terms, of world cultural heritage. Last but not least, it is also a question of international prestige that motivates the U.S.A. to be a member of these organizations. Moreover, the moral appeal underlying such motivations is often used politically in the U.S.A. in order to ensure them an influential role in these global organizations. Protection policies put forward by the UNESCO[58] do not, however, take precedence over national protection schemes (Genius-Devime 1996, p. 288).

A very important function within the area of the international preservation of monuments and buildings is held by UNESCO, which acts as a patron organization. UNESCO is a subsidiary organization of the UNO[59], which was established in 1946. UNESCO-protected buildings are administered by the World Heritage Center in Washington, DC. UNESCO aims to contribute "...to peace and security in the world" (UNESCO 1996, p. 5). By means of developmental schemes, science, and (intercultural) communication, this UNESCO project aims to build a universal sense of respect and a positive revaluation of historically relevant objects in the sense of Kant's imperative.

UNESCO's aims thus go further than protection schemes alone (Genius-Devime 1996, p. 290), and are concretized through the implementation of international standards, conventions, regulations, and recommendations. One of its larger programs, created to achieve these aims within the area of historic preservation, is the World Heritage Program. 582 cultural and natural objects were registered on the protected list in 1999. Among them are also larger zones which nominate protected areas, again comparable to the U.S. concept of historic districts.[60]

To be designated a World Cultural Heritage site by UNESCO is undoubtely the highest honor for a place, a (natural) monument or an historic ensemble. UNESCO has several U.S. locations on its list. One of the most important buildings registered on UNESCO's world cultural heritage list since 1966 is the Independence Hall in Philadelphia (World Heritage List) (Genius-Devime 1996, pp. 300). Be-

[58] United Nations Education and Scientific Cultural Organization

[59] United Nations Organization

[60] Well-known World Heritage sites in Germany include, for example, the old city centers of Lübeck (Schleswig-Holstein) and of Quedlinburg (Sachsen-Anhalt/Ostharz). Earlier known are schemes that were carried out in South Germany. Such historic preservation efforts can be followed in the journal "*Die Alte Stadt*" (published by Kohlhammer Verlag since the 1970s). Both schemes are comparable with the concept of historic districts in the U.S.A.

side such positive listings, UNESCO has set up the World Heritage List in Danger, in which endangered places and buildings are listed (Genius-Devime 1996, pp. 311). Most of the listed places are nature parks, which are threatened by intrinsic (area reduction by major fires) or system-external (interest conflicts due to tourist use) factors. Everglades National Park was added in 1993 and Yellowstone National Park in 1995.

In 1956, the International Center for the Study of the Preservation and Restoration of Cultural Property (ICCROM) was created. Based in Rome, this is an active body mediating between governments of the World Heritage Information Network (WHIN). WHIN, which is administered by the UNESCO World Heritage Center, is a so-called clearinghouse which collects information about natural and cultural property whose particular value has been already identified. The Intergovernmental World Heritage Committee has placed these buildings on the World Heritage List. The organization includes a panel of experts who give advice as to how the buildings and places registered on the list of the UNESCO should be preserved and restored.

The International Council on Monuments and Sites (ICOMOS) was founded in 1965 as a non-governmental institution with an international office in Paris. ICOMOS is a public, technical advisory institution which deals with cultural and mixed-cultural sites and provides evaluations and recommendations for the UNESCO World Heritage List for the World Heritage Committee. Further aims of the ICOMOS include stimulating public interest in the preservation and protection of monuments and buildings, uniting institutions and people who are interested in preserving monuments, ensembles and protection-worthy areas and representing their interests before international committees. ICOMOS is one of the main participants in the World Heritage Information Network (WHIN). During the first conference of ICOMOS in 1967, a set of recommendations and resolutions were adopted. This discussion paper represents a rather idealistic concept, however, and is only morally binding for the individual nations. It was not until 20 years later, in 1987, that the U.S.A. for the first time hosted this well-attended and significant conference in Washington, DC.

The U.S. national ICOMOS committee concerns itself with cultural maintenance and historic preservation on a national and international level by promoting education and further training, international exchanges of experts within the area of historic preservation, information distribution, technical assistance and documentation; it is an advisory body which organizes symposiums as well as having various other functions. In the U.S.A., ICOMOS is an organization which is independent of the government and which has as its objective in the preservation of World Cultural Heritage according to the ideas of UNESCO. At present (1999), 22 World Heritage Sites exist in the U.S.A. which have been acknowledged by the ICOMOS.[61]

Global strategies always reflect the sum of all the individual countries and their opinions about history, as well as their views concerning historic preservation and

[61] The Independence Hall in Philadelphia is one example of a designated World Heritage Site.

maintenance. Nevertheless, these organizations have tried to develop international interest and awareness, with a comparatively large degree of success as regards significant buildings, parks or cultures of universal and global value.

4.2 Federal Level

In the following, the different participants in the U.S. historic preservation process will be presented. Here, we shall differentiate between those who work with the Federal Government or operate independently of the government on a non-profit basis and those which were created and work nationally through private initiatives.

4.2.1 Federal Government

4.2.1.1 U.S. Department of the Interior

The U.S. Secretary of the Interior is responsible for the definition of the standards of all programs, which are subsequently ratified by the authorities of the individual Federal Ministries (U.S. Department of the Interior 1994a, p. 5). This also applies to state-level advisory bodies dealing with the protection and preservation of property which is or should be listed in the National Register (NR).

In 1935, the "Historic Sites and Buildings Act" was ratified by Congress. This calls on and permits the Secretary of the Interior to secure data concerning historic preservation, make statistical outlines and carry out research work, acquire property, intervene in the drawing-up of contracts ("enter into contracts") and restore buildings as well as support innovative research and develop education programs.

For many years, the role of the U.S. government concerning activities in historic preservation schemes was very passive. Among other things, this is due to the fact that it had rarely been necessary to intervene before the Civil Rights Legislation of 1964 and 1965. Furthermore, the economic potential of the preservation of monuments and buildings had not yet been identified and the cultural property of "history" had not yet aroused any social interest at that time. The unpopular topic of historic preservation had not always taken into account legal developments regarding racial equality and political correctness, which initially restricted the U.S. government's capacity to intervene.

It is only recently that the U.S. government has come to realize, as a result of scientific studies undertaken, that the preservation of buildings and monuments can itself be economically viable and represent an alternative solution in the revitalization of city center areas. Consequently, it has also become clear that historic preservation could serve the interests of other departments (U.S. Department of the Interior, HUD[62], U.S. Department of Environment).[63] For the first time, in

[62] U.S. Department of Housing and Urban Development

[63] The National Historic Preservation Act of 1966 and further tax reduction programs were implemented through law over the following years.

1969, the Department of the Interior awarded federal subsidies ("grants in aid") to the individual federal states, municipalities, groups of minorities (for example, Native-American tribes) and the National Trust. Thus, a new era dawned in which historic preservation does not exclude National Trust concerns and may also be conducive to the interests of other U.S. cabinet level departments.

In the "Standards for Rehabilitation" (U.S. Department of the Interior 1994a) the national criteria for historical buildings were established for the first time (Tunnard 1975, pp. 87). In this context, rehabilitation is seen as a process of preserving property for use through repairs or modifications. These modifications should be relevant to effective and modern functions, while the buildings' sections and features should be preserved (U.S. Department of the Interior, p. 5). The "Standards for Rehabilitation" also provide information as to how private persons and public institutions can maintain their historical properties on a long-term basis (U.S. Department of the Interior 1995).

4.2.1.2 U.S. Department of Housing and Urban Development (HUD)

The Department of Housing and Urban Development (HUD) is legislatively and administratively answerable to the Department of the Interior. HUD first mentioned the word "historic" in its legislation in 1961 (Tunnard 1975, p. 89). HUD did not provide for explicit financial or fiscal measures to protect historical buildings until 1966, although it exerts the most influence and potential for implementation in the revitalization of historical districts on a national level (Tunnard 1975, p. 89). Before 1966, only small amounts of money were made available for urban redevelopment (Uhlig 1977, p. 17).

The individual national, social and political interests at work in American cities changed after the Second World War. Prior to that, projects were promoted that focused on single objects; afterwards, whole districts began to be developed and large-scale structural renewal took place.[64] It is a disputed issue as to whether urban-social innovation could have been achieved by carrying out historic preservation work. This applied in particular to economically endangered commercial centers or road sections in the cities[65]. The redevelopment of whole districts is especially welcomed by local governments. Through revitalization programs, it was hoped that the decline in retail trade and the tendency towards suburbanization could be stopped or, at least, slowed down (Tunnard 1975, p. 89). It was not until 1966 that HUD was able to establish a legal basis for financial measures to preserve historical buildings through the Historic Preservation Act and began a national and systematic listing of historical features through the National Register of Historic Places (NR). One of the most spectacular and almost perfect textbook cases of Urban Renewal was, and still is, Society Hill in Philadelphia. Here, the preservation of the nation's historical and cultural bases aimed at reviving and promoting aspects of local life and development.

---

[64] Historic preservation did not occupy a central position during this process of revitalization.

[65] Streets, parts of streets, were dealt with by the National Main Street Program (Iv-Bos-4).

HUD realized that investment in the preservation of buildings and monuments was entirely in their interests, whereupon funds were channeled into historic preservation, the restoration of buildings and the renewal of urban districts. In 1968, the pioneering film “How Will We Know It’s Us?”, released by the HUD, served to widen the scope of social acceptance of historic preservation.

The programs of HUD show clearly that synergetic alliances can exist between historic preservation and town planning. It became clear that the social costs of programs of complete demolition in the context of the Urban Renewal Projects were too high in the intermediate and long term.[66] Among other things, national trust funds were able to be used for historic preservation, for the preservation of monuments and buildings which, after all, did much to promote existing neighborhood structures (Birch and Roby 1984, p. 203). Extensive protection schemes for districts with higher financial expenditure and authorities have, however, not been developed by the government.

4.2.1.3 National Park Service (NPS)

As early as 1872, the United States Congress created Yellowstone National Park in what were then still the Montana and Wyoming territories. The park was intended to benefit the entire population and provide them with an area where they could spend their leisure time. The establishment of this park led to an international increase in public awareness of nature protection and (natural) park systems, and was the basis for further federal government plans. It was not until 1900, however, that the government seriously began to consider the preservation of antiques and archiological sites and artifacts.

The National Park Service (NPS) was founded in 1916 by President Wilson. It was a newly-established department within the Department of the Interior’s administrative system. With the founding of the National Park Service, the administration took over nine existing historic buildings and monuments and the existing National Parks. It was not until 1933 that President Roosevelt decreed that the National Park Service take over the historic areas from other national authorities.

This also included provisions for the protection of historic war landmarks, as well as the War Department Battlefields. These had previously been administered by the U.S. War Department or the U.S. Forestry Service/Department of Agriculture. The functions of the National Park Service consist of preserving natural and cultural resources and making these NPS holdings accessible to current and future generations for the purpose of recreation and education (U.S. Department of the Interior 1996, p. 2) (Figure 4.2).

The National Park Service today cooperates with many partners which are organized on a local, state and international basis. The complex nomenclature of the National Park Service is a common source of confusion, which at the same time reflects the versatility of the National Park System. The seven main categories are as follows:

---

[66] Kapp (1950) had already drawn attention to this in his monograph.

National Park, National Preserve, National Monument, National Memorial, National Historic Site, National Seashore and National Battlefield Park (U.S. Department of the Interior 1996, p. 7). Many titles are self-explanatory, yet some are used in different contexts. The title "National Monument" is used in relation to nature preserves, historical fortresses, prehistoric ruins, major fossil sites as well as the Statue of Liberty (U.S. Department of the Interior 1996, p. 7). Due to the uncertainty of such terms, Congress and the National Park Service are striving to provide a universally clear nomenclature based on specific and consistent criteria. The legislative brief of the National Park Service must therefore be understood organically. Additions and modifications are normally made by Congress.

Apart from the aforementioned National Park areas, there are also so-called Affiliated Areas. These include a number of areas in the U.S.A. and Canada which are nationally significant properties according to the criteria set down by the park system. All areas receive technical and financial support from the National Park Service (U.S. Department of the Interior 1996, p. 99).

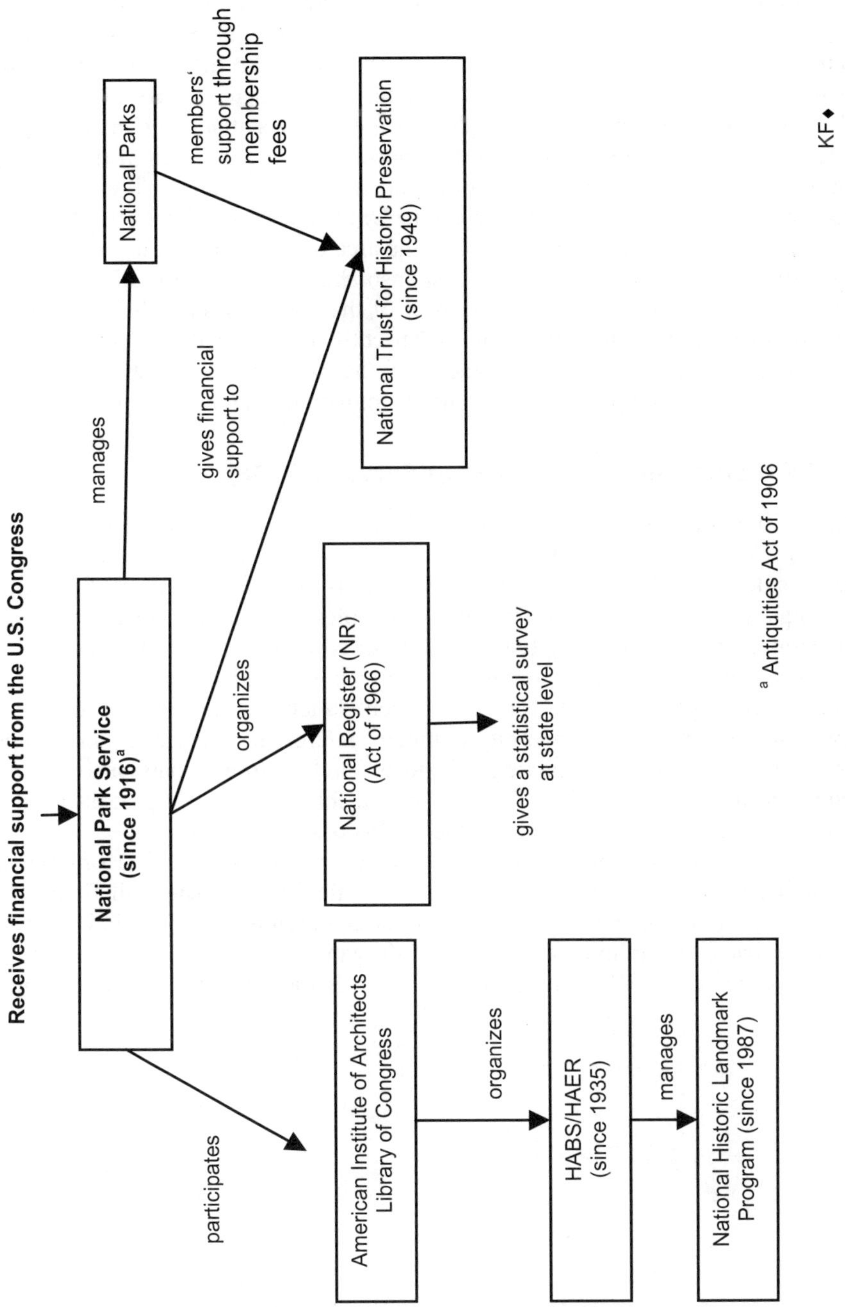

Figure 4.2. Integration of the National Park Service into historic preservation

It is possible to select an urban area such as a National Historic District through the initiative of national, state or local bodies. Through this process of official selection, the National Park Service not only allows official titles to be bestowed on areas, but also provides titleholders with certain tax benefits. Besides economic advantages for the owners, the buildings are listed statistically in the Historic American Building Survey (HABS) for shorter or longer periods of time depending on their circumstances.[67] Local governments can take further direct or indirect measures to support historic preservation. Until that point, however, all national laws apply to all of the structures within the area in question. In the case of further intervention or full local government sponsorship of the historic districts, this is generally accompanied by a legal transfer of the area to a local district. The official nomination of a National Historic District by the National Park Service, which is legislatively restricted and influenced by the existing historic districts, will be examined further in terms of their origin, administration and effectiveness.

4.2.1.4 Historic American Building Survey (HABS)

The National Park Service, the American Institute of Architects (AIA) and the Library of Congress founded the Historic American Building Survey (HABS) under the presidency of F.D. Roosevelt in 1933. This survey contains both measured drawings and photos of buildings and ensembles that are considered worthy of protection. At that time, very few U.S. architects published measured drawings or sketches of their buildings in order to make technical and structural details accessible to future generations.[68] Moreover, such sketches were rarely drawn up for new buildings. Architects' and city planners' interest in American and European history nevertheless grew, giving rise to an architectural renaissance in which many buildings were revaluated. Many called for a systematic process of stocktaking in order to enable efficient, authentic historic preservation.

HABS' aim is to gather statistics systematically and thereby to incorporate and research all possible areas falling within their field of expertise, all fields of architecture, all periods and all styles of American architecture. The individual objects are listed in archives consisting of drawings, photographs and descriptive documents. The HABS deals not only with buildings, but also with other urban features.

---

[67] In the following chapter, detailed information will be provided on the Historic American Building Survey (HABS).

[68] As opposed to the majority of European architects (Webster 1976, p. xxiii).

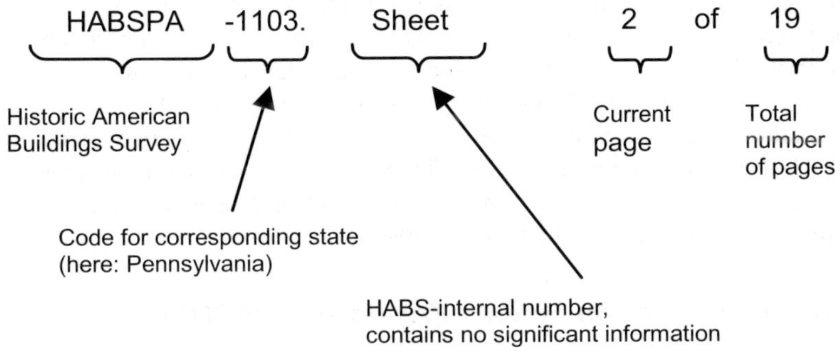

Figure 4.3. Historic American Building Survey: listing method

This program represents an ongoing process, which means that the HABS can never be seen as a complete and final list of significant American architecture, but rather one which is constantly being modified and developed. It is expected that the contents of the HABS list will constantly be extended through the collaboration of various bodies, in particular by local authorities (Webster 1981, p. xiv). The HABS itself is responsible for the quality standards of the listing, the selection of the entries and the archiving of the objects.[69]

The first four letters of the entry code assign the drawing to the HABS or HAER Program, with the following two letters corresponding to the state code. The entries are organized according to the following method. It has been shown that by revealing exception historic or cultural value, the HABS studies have facilitated the preservation of some buildings that would otherwise have been demolished.[70]

4.2.1.5 Historic American Engineering Record (HAER)

At this juncture, the Historic American Engineering Record (HAER) must be mentioned briefly. After the American Society of Civil Engineers and the Committee on the History and Heritage of American Civil Engineering designated the first National Historic Civil Engineering landmark in Maryland in 1966, there was much demand for a separate archive of U.S. American industrial heritage. Consequently, the U.S. Department of the Interior established HAER in 1969. Industrial and manufacturing-related structures had not been effectively evaluated through the documentation of relevant buildings. From an organizational point of view, HAER functions in much the same way as HABS. Like HAER, it documents, investigates and protects aspects of architectural engineering and American industrial structures. The National Park Service, the American Society of Civil Engi-

[69] The official archives of these objects can usually be found in the Library of Congress in Washington, DC.

[70] This is especially applicable for individual buildings in Society Hill, Philadelphia. (Webster 1981, p. xxxvii).

neers and the Library of Congress have reached a mutual agreement. At a time of technological change and development, it was discovered that older machines, production methods were of historical value and provided insight into aspects of technical engineering. This justified listing and protecting the objects.

The nationally-organized Society for Industrial Archeology, which works together with HAER, was founded in 1971. Their interest is above all to promote scientific studies of sites involved with industrial and mechanical processes and the production or reproduction of historical tools and machines.

4.2.1.6 Advisory Council on Historic Preservation (ACHP)

The Advisory Council on Historic Preservation (ACHP) is an independent national institution founded in 1966 under the existing National Historic Preservation Act (NHPA). This historic preservation body was created to support and encourage the protection, improvement and renewal of national historic heritage. The ACHP, which has its headquarters in Washington, DC, is the only institution which advises the Federal government on matters of historic preservation. Born (1995, p. 193) describes the role of the ACHP as a provider of federal historic preservation advice.

In 1999, twenty experts from the fields of politics and science were represented in the ACHP. The Advisory Council on Historic Preservation consists of the individual federal, cabinet-level heads whose concerns regularly involve historical property. Also present on the Advisory Council are experts from the field of historic preservation, a governor, a mayor and several citizens appointed by the President, as well as delegates from the National Trust and representatives of the State Historic Preservation Office (SHPO) (Cullingworth 1997, p. 116). ACHP membership requirements state that there must be one member on the panel of experts to represent Native Americans and Native Hawaiians respectively (ACHP 1993, p. 40).

The main aim of the ACHP is to create a forum supporting historic preservation activities on a national, state and local level, as well as the activities of the Tribal Governments. It is responsible for legislation, new aid programs and other political or economic decisions that affect historic preservation within the municipalities or on a national scale in the United States. Furthermore, the ACHP monitors and administers Section 106 of the National Historic Preservation Act (ACHP 1993, p. 7).[71]

In 1963, a series of discussions took place concerning the committee's executive power. The motive behind these was to provide an alternative to the National Register. These discussions were published in 1966 as a paper in the National Trust journal, "Preservation Today".

In 1969, the ACHP published the first guidelines for historical and cultural property included in the National Register. Buildings whose historical or structural features were not yet recognized by the state were also included on this list. The committee was particularly interested in historic preservation in the U.S.A. as a

---

[71] For a discussion of "Section 106", see next section.

way of enabling (new) discussions of locally significant buildings within the individual states. Locally important objects or those worthy of preservation are not included in the national protection programs of historic preservation, as has been pointed out previously (ACHP 1993, p. 7).

The functions of the ACHP are, in fact, only administrative; the body has no legal power and occupies a purely advisory position. Nevertheless, the ACHP has become an important resource for public and political bodies involved in historic preservation. An important goal of the ACHP is also to promote public interest in and awareness of the different levels of education and the way in which plans are executed (see ACHP 1988, 1992 and 1998). In this context, the ACHP regularly awards money and grants in the name of the President, in order to ensure the execution of successful historic preservation work (ACHP 1998 and 1992).

4.2.1.7 A Legislation of Historic Preservation: Section 106

In the following, an important statute from the domain of historic preservation will be described in more detail. This statute is of importance in the empirical section and is of particular significance when examining the districts studied here. Furthermore, these laws are substantially different from those existing in Germany and Europe. Section 106, which is administered by the Advisory Council on Historic Preservation (ACHP), is an important provision of the laws passed in 1966 as part of the Historic Preservation Act (NHPA) (see: Public Laws 89-665; see also the discussion in: NTHP 1997, p. 54).

Section 106 of the National Historic Preservation Act requires Federal agencies to take into account the effects of their undertakings on historic properties and afford the Council a reasonable opportunity to comment on such undertakings (Tunnard 1975, p, 88). Should modifications be identified, the issue must be submitted to the Advisory Council of Historic Preservation for inspection. In this way, representatives can act to ensure that state projects, listings, and advisory offices work alongside the individual private restoration plans, and not against them.

A listing in the National Register at the same time means that the object is protected under Section 106. Moreover, protection can be extended to objects that qualify for inclusion in the National Register.

The State Historic Preservation Officer (SHPO)[72] holds the highest executive power with respect to Section 106. The latter is of particular interest in relation to the holding of private property in the U.S.A. Intervention in private ownership rights is regarded by private concerns as one of the largest obstacles within the area of historic preservation. Such obstacles are dealt with to a large extent by this legislation. Federalism is a deep-rooted mindset in the U.S.A. and can hinder plans for historic preservation.[73] Section 106 makes it possible to override a decision reached by the individual states.

---

[72] This also includes the Tribal Historic Preservation Officers (THPO), who do not figure in this analysis.

[73] See glossary for a detailed political and administrative definition of this term.

An interesting and highly controversial issue discussed in the context of Section 106 was the Vieux Carré case (the Old French Quarter in the old part of town) in New Orleans. The district had already been nominated as an historic district in 1965. The U.S. Department of Transportation wanted to construct Interstate 310 in such a way that it would have affected the historic district. The National Register allowed citizens to object to this road construction because the traffic system in the Vieux Carré would have affected residents, visitors and tourists. They would have failed to derive any benefit from the building of the new highway. Furthermore, the visible highway would have ruined the townscape and deterred future investors in the commercial sector. Although there were many obvious and comprehensible reasons for preventing the construction of this highway, the highest city building authority had had spoken out in its favor. After further debates and consultation with the ACHP, drawing on Section 106, I-310 was not built, because public need for the highway was not great enough to justify the destruction of the French Quarter (Tunnard 1975, p. 97).

The U.S. Supreme Court created another unusual precedent in 1978. In the case of Penn Central Transportation Company vs. City of New York, this highest U.S. judicial body overruled all other decisions and denied permission to construct an office building on top of the Grand Central Terminal. This decision in favor of historic preservation interests represented the first major preservation case dealt with by the Supreme Court and is an important event in the context of Section 106.

In other cases, however, decisions made by the ACHP in connection with Section 106 merely delayed matters and tested public opinion (Birch and Roby 1984, p. 202). Social and economic constraints continued to affect the outcome of such cases significantly. More information concerning this discussion can be found in Attshuler (in: Krueckeberg 1997, pp. 190).

To summarize, it may be said that this is one of the few cases in which a Federal office can be influential in making decisions and can affect existing decisions by invoking Section 106.

4.2.1.8 National Register of Historic Places (NR)

The major aim of the National Park Service is to preserve historical property, and its role in U.S. American heritage reflects national interests (National Park Service et al. 1994, p. i). This gained particular significance when new legislation was passed in Washington according greater importance to the preservation of monuments and buildings, including the Antiquities Act of 1906, the Historic Sites Act of 1935 and the National Historic Preservation Act of 1966. The aim is to coordinate collaboration between public and private initiatives in the area of historic preservation (National Park Service et al. 1994, p. vii).

The National Register of Historic Places (NR) was established through the National Historic Preservation Act of 1966, with the first list being published three years later. The NR's function is to list America's cultural resources (Cullingworth 1997, p. 116). Previously, only an incomplete list existed, including some historical features in the National Park System and a few National Historic Landmarks. In the same year (1969), a provisional, non-binding list (Advisory List) was issued by the ACHP, in which further objects were proposed for inclusion in the National Register. It was not until 1994, 25 years later, that an updated list with all recent inclusions was published (National Park Service et al. 1994). To date, about 70,000 objects have been listed in the National Register (ibid. p. viii).[74] All objects situated within the National Park System are listed. Moreover, objects situated outside the National Park System, nominated by governments, private organizations and individual persons, have also been added to the National Register. It is the individual states, however, which are responsible for ensuring that buildings are listed, with cities responsible for systematically listing features of interest (Birch and Roby 1984, p. 202). This ensures that the direct costs of the NR are kept relatively low.

The National Register contains the complete official list of recognized listed objects and properties, including houses, historic districts and other structures. The list is administered and extended by the National Park Service on behalf of the U.S. Department of the Interior. The National Park Service developed the criteria which determined which buildings may qualify for listed status on the NR. These criteria apply to every object on the list without exception (National Park System et al. 1991). The NR documents the physiognomic structures and significance of districts and regional areas such as neighborhoods, places, buildings, historic districts, individual structures and other objects that play a significant role in the prehistory and history of the United States. These objects represent local, state and national history and experience (National Park Service 1994, pp. vii). "In addition, the National Park Service has developed criteria for the recognition of nationally significant properties, which are designated National Historic landmarks and prehistoric and historic units of the National Park System" (National Park Service 1991, p. i).

The systematic listing of the objects in the National Register is generally regarded as contributing much to growing national awareness of the need to preserve monuments and buildings. Historic preservation has thus been drawn to the attention of society and has become socially acceptable, as a result of which, objectives and opportunities can be discussed anew. In addition to key social concerns, such as patriotic and aesthetic issues, discussions began to focus on economic and social questions. It has since become widely accepted that the population has a responsibility to identify historical buildings, designate them as such and preserve them. This is due in no small part to the emphasis on individual well-being when implementing historic preservation.

[74] Of a possible 900,000, according to some estimates (Iv-Phl-10).

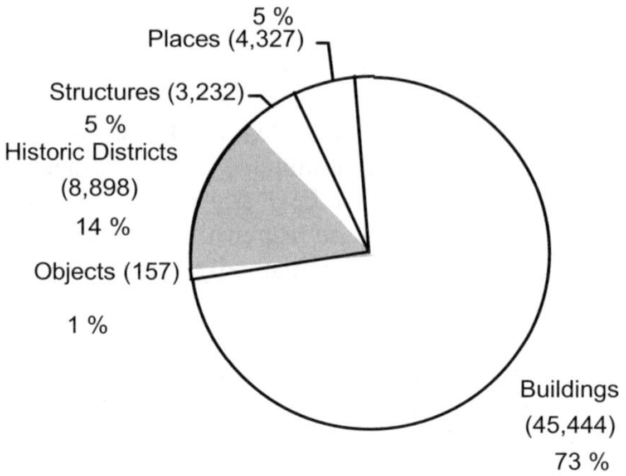

Within the *historic districts* there are approx. 1,000-2,000 individual objects. This figure cannot provide information about the total number of objects registered by the National Register.

Figure 4.4. Object types in the National Register (Source: National Park Service and National Trust for Historic Preservation 1994, p. vi)

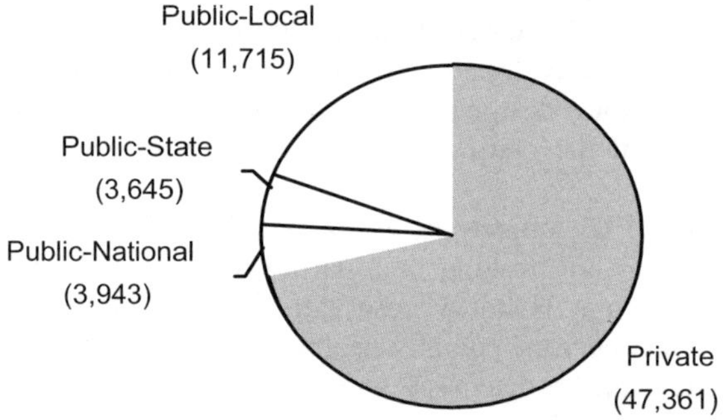

Some objects are owned by more than one person/institution.

Figure 4.5. Distribution of ownership of objects listed in the National Register (Source: National Park Service and National Trust for Historic Preservation 1994, p. vii)

The advantage for owners whose property is listed in the National Register, besides the honor of a nomination and the additional plaque that accompanies it, is that the public planning processes will take the listing into consideration, offer them substantial tax reductions and qualify them for public programs promoting endangered objects. The owners of properties that are listed in the NR are eligible for an investment tax reduction of 20% if they carry out restoration work on the building that has met with public approval. Tax benefits are also granted for donations that support conservation measures.

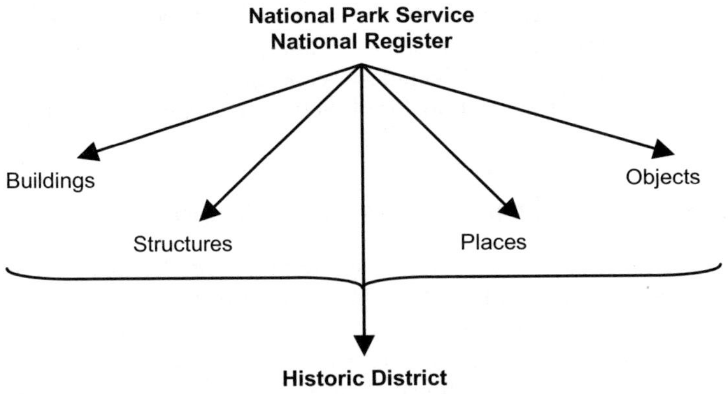

Figure 4.6. Objects included in the National Register

As long as no public money is donated towards the maintenance of a building, the owner can preserve, treat or use his property much as he chooses. This clause only becomes void if local governments list buildings locally, for example, by setting up an historic district, in which case the owner has to accept the local laws that apply to that historic property.

The National Register does not take into account cultural events that have no distinctive physical features. The listings only include physical and concrete property units that are (relatively) fixed in their environment. In order meaningfully to systematize the National Register, the individual objects in question have been divided into groups as indicated below. Objects do exist that could be classified within several categories; thus common sense must be used to determine effective classification (National Park Service et al. 1994, S.4).
The nomenclature designates NR *buildings*, *structures*, *objects*, *sites* and *districts* (National Park Service et al. 1994, pp. 4).

Examples of buildings are private homes, government buildings, horse stables, city halls, skyscrapers, post offices and schools. The term "structure" is used to differentiate a particular object from a building. It defines all objects that are not used to house people but are still functional. Examples of this form of object are: bridges, canal, dams, fences, highways, lighthouses, silos, tunnels, and windmills. Objects are constructions that are essentially aesthetics and whose space within their structural environment is relatively small and insignificant compared with the whole. Objects can, for example, be sculptures, fountains or statues. A site will be included in the National Park Service list (1994, p. 5) if is a place where an historical event occurred. In most cases, it does not matter whether the place still exists in its original condition or whether it has disappeared or fallen into disrepair. This can be the case when dealing with places that have prehistoric or archeological importance. Examples include be battlefields, encampments, landscapes and old tracks and trails.

The most important and regionally relevant term for this study is that of the historic district. "A district possesses a significant concentration, linkage, or continuity of sites, buildings, structures, or objects united historically or aesthetically by plan or physical development" (National Park Service et al. 1994, p. 5). An historic district therefore (usually) includes a concentration or combination of all categories of structure that the National Park Service has documented its National Register.

4.2.1.9 National Historic Landmark Program (NHLP)

Since 1960, the National Park Service has administered the National Historic Landmark Program on behalf of the U.S. Department of the Interior. This serves to preserve processes of identification, recognition and nomination of certain locations, (non-structural) locations and objects of national historic significance. The legislation that underpins a National Historic Landmark (NHL) is recorded in the U.S. Code of Federal Regulations (36 CFR 65). Such landmarks are considered to be of national importance. Buildings whose significance is purely local are excluded from this program. All objects reflecting U.S. history should be included. The NHLP helps governments, independent private organizations and individual persons by placing public means at their disposal and offering assistance through nomination as being worthy of protection. Moreover, it identifies and nominates many places that are not included in the National Park System. By 1987, 1,811 objects had been designated. The designation process is based on the results of studies carried out by the National Park Service proposing objects to the U.S. Department of the Interior for designation. The final decision as to whether the property is listed is reached by the Secretary of the Interior (U.S. Department of the Interior 1987a). At this time, the U.S. Departent of the Interior has listed about 2,300 objects. Constant additions are being made to the list.[75]

National significance is accorded to districts, places, buildings, structures and objects of exceptional value or quality and which illustrate and facilitate the interpretation of American heritage historically, architecturally, archeologically, mechanically or culturally (U.S. Department of the Interior 1987a, p. ii).

Once a building has been incorporated into the National Landmark Historic Program, the owner receives a nomination document that permits him to put up a plaque and thus make the nomination public. The owner does not, however, lose the right of ownership of his property, and U.S. Department of the Interior does not repossess the property, nor is the owner obliged to make his property accessible to the public.

A number of other advantages are offered to the owner in addition to the official listing of his property, among which is the opportunity to apply for assistance from support programs and thereby to achieve higher social and idealistic status. Along with nomination and recognition on a national level, it is, however, possible for local and state governments to place further limitations on the buildings. In the case of improper maintenance or a loss of the reason for nomination, the owner's Landmark Designation can be withdrawn. In this context a study has been pub-

[75] See www.cr.nps.gov/nhl (11th May, 2000).

lished that lists damaged and endangered objects (U.S. Department of the Interior 1994b). According to this study, 17% of all National Historic Landmarks are damaged or seriously endangered, while 7% are in a critical state (ibid. p. 111).

Since 1966, new legislation has been passed which makes the National Historic Landmark Program subordinate to the National Register. The National Park Service is responsible for its administration. The program is directed and supervised by the State Historic Preservation Officers (SHPO). The SHPO is also charged with making nominations (Birch and Roby 1984, p. 202). The costs of implementing such programs are generally passed on to individual state authorities (much like the National Register).

4.2.1.10 National Trust for Historic Preservation (NTHP)

The National Council, the predecessor of the National Trust for Historic Preservation (NTHP)[76], was founded as early as 1947. The U.S. Congress then extended the National Trust in 1949. Even today, it represents the only national, private, non-profit organization that has made it its duty to encourage the participation of the public in preserving historical places, buildings or ensembles and objects that are significant to American history and culture. The NTHP is supported through contributions of the membership organizations and funds from the National Park Service. The laws developed over many years and the various programs of the National Trust for Historic Preservation have, for the most part, grown along with the needs of historic preservation.

As a result of the Mayors' Conference and the National Trust for Historic Preservation, a manifesto entitled "With Heritage so Rich" was produced and published in 1966 (NTHP 1983). It contains, above all, explanations of the historic preservation legislation and recommendations for citizens (Cullingworth 1997, p. 116).

Today, the National Trust has 250,000 members and seven regional offices, opened between 1971 and 1977. Furthermore, the National Trust presently owns 18 museums and historical houses, some of which have been restored by the National Trust. The National Trust advises and works together with over 1,000 regional groups in all of the 50 states. The organization still works under national sponsorship. 90% of the budget comes from contributions and donations by members and independent sponsors.

The responsibilities of the National Trust are not of an executive nature; instead, it performs duties such as carrying out tours in the historic districts, issuing recommendations for preservation work and acting as a national advisor in matters of historic preservation. The main aim of the National Trust is to acknowledge and understand the plurality of American cultural heritage. This revitalization is intended to promote the vitality of communities through National Trust protection of the historical American landscape.[77] To achieve this, the National Trust has long supported heritage education, neighborhood revitalization, the development of or-

[76] Often abbreviated to "National Trust".

[77] The term "landscape" is used metaphorically and not in its regional geographical sense.

ganizations and other programs.[78] In 1973, the National Trust organized and sponsored the journal, "Preservation Week", which brought together national, state and local interests for the first time.

In 1980, the National Trust created the Main Street Center Program. The purpose of this program was to revitalize the old main shopping streets that defined the economic nucleus of cities through special administrative, consulting and financial means. Targeted programs enabling joint ventures with communities allowed this program to strengthen a number of communities and neighborhoods. One year later, the National Trust established an additional Inner City Ventures Fund, the aim of which was to support historic preservation activities in neighborhoods populated by low-income groups with especially high proportions of minorities.

The National Trust thus has a dual mandate, to recognize and conserve the various characters and cultures of America's cultural heritage, and to ensure the preservation and revitalization of the communities. In this respect, the National Trust plays a leading role in guiding communities. For this reason, the scope of the National Trust has been extended to strengthen its influence in the education sector, neighborhood revitalization has been encouraged, the service areas of organization development expanded, and training of experts and land conservation strengthened (see Copps 1995). Since 1979, the Trust for Public Land (TPL) has worked together with the National Trust. These institutional joint ventures preserve historical buildings, archeological sites, and landscapes. Copps' (1995) monograph, "Views from the Road", describes the results of this cooperation between the National Trust and the TPL. The goal is to help communities on a local level to discover and protect their natural and historical resources.

4.2.2 Private Partner Organizations

4.2.2.1 American Institute of Architects (AIA)

The American Institute of Architects (AIA), founded in New York in 1857, also plays a significant role in the context of the American preservation of historic buildings and monuments. The AIA is a private national organization with headquarters in Washington, DC. Any registered architect or student of architecture may become a member. Special membership categories exist for persons in the allied disciplines, such as urban and landscape planning. In 1994, the AIA had 56,000 members, of whom only 10% were women, with 7% belonging to minorities. An education or previous educational background in architecture is not necessary for a general membership. An architect may, however, only list himself as an AIA architect after passing a comprehensive examination. The AIA has established its own individual quality standard in the field of architecture by offering courses and additional tests for distinguishing AIA registered architects (Iv-Phl-9).

[78] www.nthp.org.

The first American Architect Licensing Law was passed in 1897 by the state of Illinois. It was not until 50 years later, in 1951, that the last two states, Wyoming and Vermont, finally acknowledged these laws. Each state has its own chapter, which again is divided into regional sub-groups.[79] Each level in turn has its own special interest groups (division). In 1932, the at that time Europhile and Anglo-Saxon AIA initiated a systematic scientific and historically correct inventory using measured drawings.

Many initiators of the AIA had received at least part of their education in Europe, which meant they were familiar with European precision and systematic work using measured drawings and inventorizing of objects. Due to war circumstances or other reasons, it also happened that a number of professors emigrated from Europe and taught at influential universities, such as the Massachusetts Institute for Technology (MIT), the University of Pennsylvania or Yale University.

The regional involvement of the individual departments of the American Institute of Architects differs and have a varying influence on historic preservation decisions made on a national, local and state level. At regular intervals, generally once month, current planning decisions by the City Council are discussed and summaries and results of the public board meetings discussed. In controversial and difficult individual cases, the AIA Division drafts an official opinion or counter-proposal, which is presented at an appropriate forum (media, hearings). Most of the time, however, public offices (city administrations) solicit advice from the AIA. In addition to consultation concerning individual objects, advice is also sought regarding ensemble protection by public authorities.

Many members of the American Institute of Architects are also represented on the local Boards of Architectural Review (BAR)[80] in a consulting capacity. In matters of new decisions they are an important committee that defines the direction of a position (Southworth and Southworth 1992). In some cities, the Historic Preservation Divisions of the AIA have influenced the architectural guidelines for the historic districts (ibid. 1992).

Many architect's offices, especially in cities with a large number of historic buildings and extensive historic artifacts, have recognized that the architectural preparation of houses in the historic district is a lucrative business. It is thus not uncommon to see specialization in historic preservation in cities like Philadelphia and Boston.

The Boston Society of Architects (BSA) is an example of a very active group of the AIA parent organization. The BSA is situated in the east of Massachusetts, and, with around 2,000 architects and 1,000 associate members, it is one of the largest subdivisions. In addition to lectures, educational programs and so on, they have also written extensive scientific works about the Boston SMSA which well depict Boston's architectural inventory (see: BSA 1992 in: Southworth and Southworth 1992).

[79] In Philadelphia and Boston, the AIA chapters were established as early as 1887.

[80] Many other names may be found, such as the "Board of Architectural Review" and the "Architectural Commission".

4.2.2.2 Daughters of the American Revolution (NSDAR)

The Daughters of the American Revolution (DAR)[81] play a significant role in the historic preservation of America (NSDAR 1996, p. 9). Although this is a national organization, its influence on the preservation of buildings and monuments in the America is only regionally distinctive and often reflects the initiative and involvement of an individual person. In some cities such as Charleston, they have contributed significantly to the preservation of historical structures (Chase in: NTHP 1998, p. 7). These important and, in some cases, even pioneering activities apply especially to the first phase of historic preservation, from around 1900 to 1930. At that time, historic preservation work was more nationalistic and patriotic, with ideological motivations: the main objective was to keep the colonial structures of the cities from deteriorating and to preserve their remaining structures for educational purposes for present and future generations.

The DAR was founded as an independent non-profit organization in 1890; in 1896, it was incorporated by Congress. Their goals include historical research with publication of the results, support of education and enlightenment, and patriotic preservation of the ideas of post-colonial times, i.e. the beginnings of the United States (NSDAR 1996, Article I, p. 9). Topics which concern the members of the DAR mainly deal with the colonial period, the Civil War, and the early republic (NSDAR 1996, p. 9). Hunter describes the early features and motivations of the NSDAR in her monograph (1991, pp. 127).

Since 1940/1941, the DAR has had its own library, an archive and a museum in Washington, DC, in which it makes its own research, acquired and donated objects, data and sources accessible to the public (Hunter 1991, pp. 10). It is a private and exclusive organization, which Europhile tendencies traits as shown by its individual activities, methodology and approaches. The first systematic inventorying of the objects owned or administered by the DAR was not commissioned by the organization until 1979 (NSDAR 1979). Barrington (1941) had attempted an inventory in the form of a monograph on the occasion of the 50th anniversary of the NSDAR, but insufficient data made it incomplete. Conspicuously missing was data on tracing of individual objects, which the author had ommited.

Membership requirements for the DAR or one of its regional groups are very strict. Every member has to be able to prove her direct descent from an ancestor who contributed to achieving American independence. Any woman over the age of 18 can apply for membership by showing evidence of appropriate background. There is a long, detailed list documenting the acceptable contributions to America's independence. This might, for example, include having signed the Declaration of Independence, membership in the military service or in civil service (NSDAR 1996, Article II, Section 1-3, pp. 10).

[81] In order to avoid confusion, it should be pointed out that the National Society of Daughters of the American Revolution is often referred by the abbreviation DAR.

4.3 State Level

4.3.1 State Historic Preservation Office (SHPO)

Since 1980, it has been legally established that every state must have its own State Historic Preservation Officer with an office at his disposal. The executive and judicial functions are, however, different from Germany: The State Historic Preservation Officer is appointed by the respective state. The State Historic Preservation Officer is the highest office regulating, historic preservation activities state, and determines even the political course of the state regarding use of funds and activities in the field of historic preservation.

The National Conference of State Historic Preservation Officers was created in 1969 to provide an historic preservation forum for the individual states, in which information and solution strategies are regularly exchanged regarding current historic preservation questions and problems (tourism and preservation of monuments, as well as local concerns such as earthquake safety).

The task of the State Historic Preservation Officer is to examine the procedure of historic preservation on a state level and to inspect the work after it has been completed. This applies especially in the context of the Tax Reform Act of 1976 to tax breaks for buildings listed in the NR or located within an historical district (Gutschow 1978, p. 134). The basis for this examination is the "Standards for Rehabilitation" (U.S. Department of the Interior 1994a).

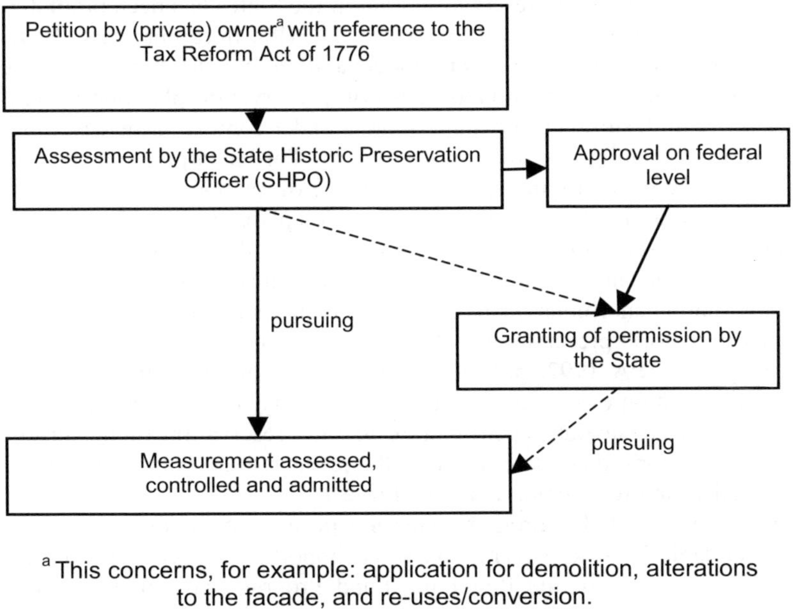

Figure 4.7. Responsibilities of the State Historic Preservation Officer (SHPO) in connection with the Tax Reform Act (Source: based on Gutschow 1978, p. 134)

The State Historic Preservation Officers have to account for their activities to the U.S. Department of the Interior. Furthermore, they are required to establish a State Historic Preservation Review Board and to set up a plan for ensuring of active public participation in the state's historic preservation program (NTHP 1997, p. 54). A few federal states distribute so-called catalogs with the HABS/HAER listings which make clear that the state as local authorities are to continue this national list. This effort, however, depends much on the political significance attached to historic preservation and on the budget of the individual state.

Finally, the State Historic Preservation Officers act as Federal functionaries of historic preservation. There, they occupy a mediating position between state and local government. Often, they are only called on to reduce the historic preservation budget or to forward program changes to the appropriate offices. The capitals of American states are often small cities, situated at some distance from the large metropolitan areas.[82] The individual cities states should not be treated differently. Nevertheless, planners in the main cities often have little direct interest in the core cities, which leads to an administrative detachment concerning the historic districts. This diminished interest has a clear effect on the intensive support of historic districts in the framework of revitalization programs, for example. The SPHO also holds executive power for legislation in the context of Section 106.

4.3.2 Historic Preservation Planning

For some years, the Historic Preservation Plan in the States has been promoted on the Federal level. This program was created by the NPS. In recent years, the focus of the states has gone beyond the mere understanding and statistical recording of historical and cultural resources. Good Historic preservation planning should incorporate above all historic preservation in the broader context of the development plan, as well as planning jurisdiction (Iv-Har-1). This incorporation documents simultaneously the generally increasing acceptance of methods for profitably integrating the preservation of monuments into complex plans (Iv-Phl-10).

The historic preservationists are concerned above all with the benefit to the historical substance. The benefit to both interested parties arising from the historic preservation plan is enormous: plans enable interactions and cooperation between participants in historic preservation, who previously pursued various smaller, individual projects (APA 1992, p. 33). An important example is the so-called revolving fund. Creation of the legal situation by the city in the planning and fiscal area represents a considerable contribution to the preservation of the historical substance. The historic districts covered in the present study are also components of the State Historic Preservation Planning Projects.

In the framework of this program, Georgia raises two important issues concerning sustainable historic preservation as an important aspect of planning: On one hand the extension of the Revolving Fund Program, and on the other, in-

---

[82] For example, Harrisburg, PA and Columbia, SC. Exceptions to this rule are Boston, MA and the fast-developing urban region of Atlanta, GA.

creased awareness for catastrophic prevention measures. In this regard, the state mainly provides many usefull information.[83]

4.4 Regional and Local Level

The various local organizations are summarized in the following.[84] This concerns more the actual initiators of historic preservation work in the historic districts of America than external influences controlling the preservation of buildings and monuments.

4.4.1 Local Historic Preservation Office (Historical Commission)

Due to the prevailing federalism[85] in the United States, the American historic preservation offices are equipped with varying executive authority, and the number of employees in the individual cities, as compared with the population of the cities, varies greatly. The appointment and administrative classification of the offices also varies considerably from state to state.

Every city has also a specified time limit, according to which a house or building may be deemed historical. Boston specifies a period of about 50 years, whereas there is currently no limit in Philadelphia.[86]

Table 4.1. Classification and appointment of historic preservation offices (Source: Interviews with the relevant City preservation officers)

City	Name of the historic preservation office	Employees
Philadelphia	Philadelphia Historical Commission (own office, reports directly to the City Manager)	6
Boston	Boston Landmarks Commission (in the Environment Department)	About 15[a]
Charleston	Special Committee on Zoning (in the Department of Planning and Urban Development)	About 10[a]
Savannah	Savannah Historic District Board of Review (own office, reports directly to the City Manager)	2

[a] The number of employees for Boston and Charleston are estimations, as employees in some cases work in several departments.

[83] For this see: www.dca.state.ga.us (11th May, 2000).
[84] For a discussion of these organizations, see also: Murtagh 1997.
[85] See glossary.
[86] In Hamburg (Germany), the time limit is currently 30 years.

Table 4.2. Current time limits: minimum age of historical buildings (selection) (Source: City historic district (zoning) ordinances, 1999)

City/Institution	Minimum Age
Philadelphia	No official limit, all buildings listed in the Philadelphia Register are considered "historic". An additional rule exists: no building constructed by a living architect can be considered under existing historic preservation laws
Boston	50 years, or exceptional designation by the Boston Landmark Commission (BLC).
Charleston	All buildings are at least 75 years old or are listed in the Historic Inventory Map of Charleston city.
Savannah	No limit specified: all buildings published in the HSF - Historic Savannah Foundation (1979) – or which were subsequently listed by the mayor.
National Trust for Historic Preservation (NTHP)	50 years

All state-recognized places and houses listed for historic preservation are then subject to legal ordinances (historic district ordinances or historic zoning ordinances), recommendations, and separate tax reduction. The goals of the local state organization are, first of all, the preservation and conservation of the cultural assets in the corresponding districts. Artistic and aesthetic questions are of secondary importance: historic preservation does not always have to look "good". This subjective influence is emblematic of the conflicts of interest between tourism, the economy and the local interests of the residents of an area.

Each City Historic Preservation Officers basicially views himself/herself as the leader of the National Register on the local level (Iv-Bos-1). The legislative authority, however, is different in every city and every state. How active the historic preservation program is in a particular city can also be seen from the way the four categories of new construction, demolition, addition, and expansion are handled (APA 1992, p. 17). Some of the historic districts have building codes for all categories, others only for a few (Appendix B.3 and B.4).

The laws and maxims of the state are, for the most part, assumed by the regional offices. Often there are regional subtleties and, above all, different interpretations as well as different degrees of accuracy in checking these laws and specifications and properly implementing them. Likewise, the economic components of historic preservation diverge considerably in the four cities examined here.

Because of the federalist nature of the system of states in America, it is important to bear in mind that a building that has been deemed historic by the city may not necessarily receive such recognition from the state. This leads to a situation in which the owner of these buildings has to find other financing means to maintain the structure. Generally, the cities have more local or regional historical monuments and constructions listed than the state. Almost all listed historical structures listed by the state are, however, also recorded in the local register.

Public space in the city is likewise subject to the territorial control of individual groups and therefore the source of conflicts of interest. Consequently, in some public spaces, certain social groups are not welcomed. The state and the city administration are at a loss when it comes to planning such public places. The political and administrative unit of the state must take into account the diametrical expectations of the residents: on the one hand, the possibilities for a specific lifestyle for the middle and upper class and, on the other hand, the need for affordable housing space and access to collective consumer goods.

The local historic preservation office is to be understood as the local organization for the National Register. For the most part, they administer both this nationwide monument file as well as a local inventory list, which at the same time represents the basis for the city archive. Also included in this context are the coordination of commissioned studies and the creation of building codes or design guidelines. The Historical Commission furthermore carries out more or less active inspections that verify adherence to historic district ordinances (Iv-Phl-16) (Figure 4.8).

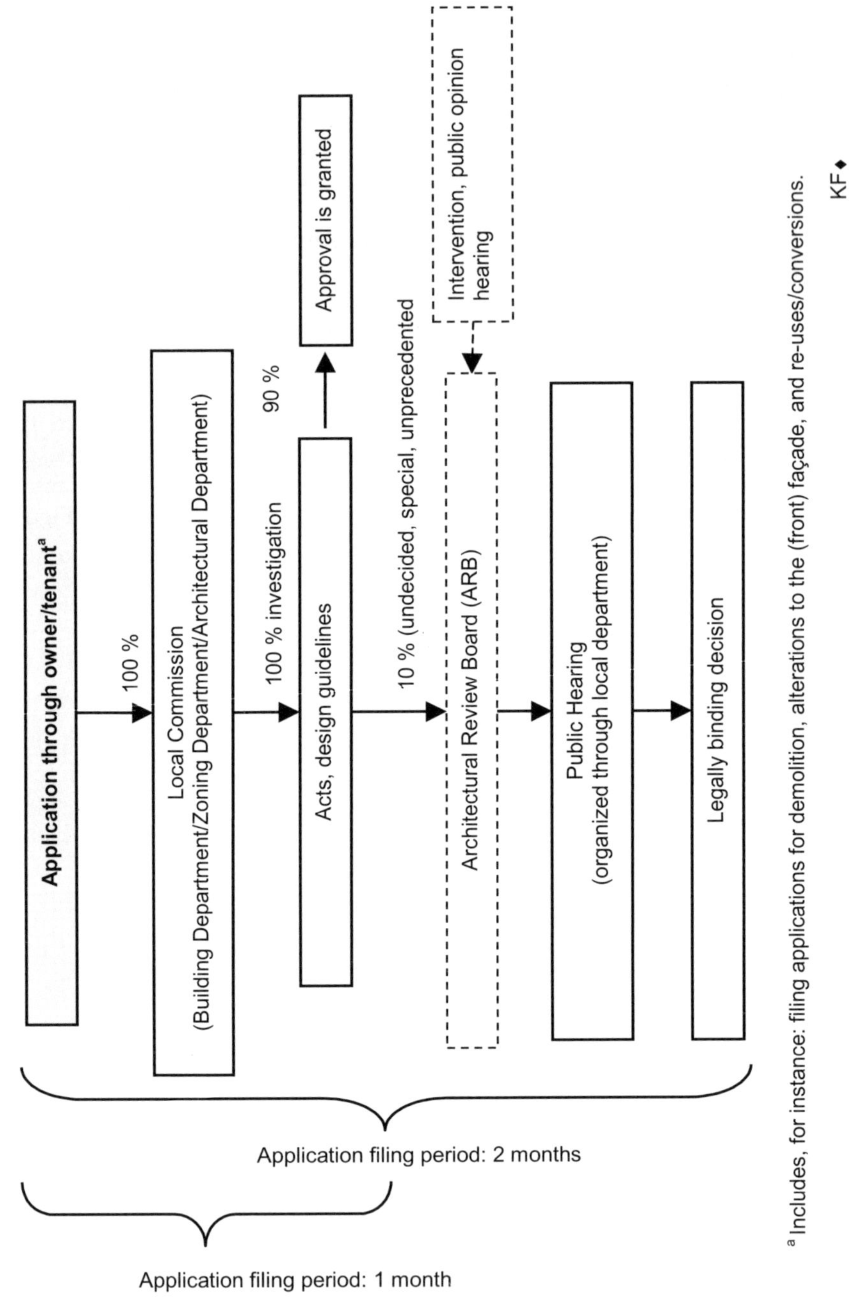

Figure 4.8. Filling out an application form for building alterations within the historic district

The office organizes the administrative network for the existing historic districts. This includes organizing public hearings and appointing an appropriate commission such as the Board of Architectural Review (BAR). The Historical Commissions, have a greater or lesser influence, depending on the staff and the extend to which they cooperate with the other local entities. Some are more active by declaring new city areas to be historic districts, while others play a more passive role (Iv-Phl-11, Iv-Bos-1).

4.4.2 Private Organizations

In the United States, the respect for and the interest in ancient relics, antiques and historical structures are more an individual than a civic duty, as opposed to Europe (Hosmer 1965 and 1981). It is for this reason that many organizations have been founded that make historic preservation their objective. Consequently, historic preservation has to be seen more as a private initiative with private interests than is the case in Germany. In the following, the participants will be introduced and their influential activities in the historic districts explained.

4.4.2.1 Preservation Alliances

Following the growth of general interest on the part of public and private entities in historic preservation, the Preservation Alliance of Louisville and Jefferson County in Kentucky was founded in 1972. The organization represented the first city and SMSA to bring private organizations, neighborhood organizations, Preservation groups and Historical Societies together. Problems concerning historic preservation are brought before these organizations. This public-private form rapidly became a model for many other cities and communities. The Preservation Alliance finances itself through membership fees and donations from individual persons who, in turn, can derive tax benefits. The Preservation Alliance can be seen as patron of the local private participants in the field of historic preservation.

The Preservation Alliances help with applying for funds to support institutionalized or private measures of preservation. They also provide pro bono consultants who, while they are politically independent, do in fact play a rather political and subjective role in the individual cities. Their authority in historic preservation and qualifications in this field are very high (Iv-Sav-1, Iv-Bos-3). In the cities that are the subject of this study, it becomes clear how different the regional structures of this organizational form are and what influence it exerts.

Depending on budget and involvement, the Preservation Alliances organize so-called revolving funds. They purchase historically valuable houses, which they renovate or, under certain conditions, sell to buyers who are willing to do the restoration work. The houses are then offered for sale (Iv-Phl-15, Iv-Sav-2, Iv-Bos-3).[87] Although the revolving fund method is still being practiced, the housing po-

[87] Hofmeister (1977, p.146) describes in great detail the methods and concepts of the revolving funds in Savannah.

tential is depleted. In many cases, it is difficult to resell the houses, and the empty houses are a source of aggravation to the neighborhood organizations. Demolition for reasons of safety is often preferred by the neighborhood organizations (Iv-Phl-8).

4.4.2.2 Historical Societies

Historical Societies in the context of the private participants are organizations that joined together at the beginning of the 20th century to address issues of historic preservation. This early phase of historic preservation was mainly patriotic in nature and dedicated primarily to the preservation of individual objects. These were often museums or houses. Membership requirements for these societies varied, but the trend in the beginning was kept rather elite (Iv-Bos-3, Iv-Bos-4). A certain class affiliation or nomination by other members was mandatory. Later on, it became possible to join the Historical Societies through membership fees. The specific contributions to the preservation of monuments were often not historically authentic, which was partly due to the fact that information about the building was lacking. During the 1930s, the historic preservationists could rely on the measurement drawings and historical research of the HABS.

Since the Historical Societies represented, to a large extent, a group of rather well-to-do traditionalists, their influence differs regionally and locally. These private organizations are not small layers of the government, but independent organizations. Many, however, were founded under Federal legislation; consequently, they pay no taxes and therefore are obliged to operate as charitable organizations.

The Historical Societies see their main tasks as similar to the Preservation Alliances, namely in the public moderation of historic preservation, the initiating of projects supporting collective interests in neighborhood assistance and in the support and establishment of contacts with interested people and participants as well as in the mobilization of citizen collaboration.

Often, the Historical Societies rent, lease or buy buildings using the revolving funds method, to rescue, protect, or restore these structures. To find new uses for historical buildings is the main occupation of these groups (see Costonis 1974). Nowadays, most of the historic districts have already established, structural and legal foundations. In these historic districts, the task of the Historical Societies is, above all, the obtaining of support and research into the historical background as well as in the function of inspecting in the sense of historic preservation. When historic district ordinance are contested, they also partcipate in court cases as a joint plaintiff or offer their juristic assistance to the neighborhood organizations (Iv-Phl-15).

4.4.2.3 Neighborhood Organizations

One of the most important private motivators in historic preservation on the local level is the neighborhood organization.[88] Even though the body of data concerning the work of this type of group is rather unclear, their contribution is by no means marginal (Mayer in: ILS 1999, p. 14). When assessing the outstanding significance of the neighborhood as a social and territorial subunit as well as a source of identification for residents of a city, Philadelphia has been repeatedly cited by social scientists, historians and politicians (see: Weiler 1974, Cybriwsky et al. 1986 and Cybriwsky 1986). Neighborhood organizations have not been only the result of protest movements, and representatives of the neighborhood organization do not always see themselves as opponents of the public sectors.

We are not dealing here with a formal unit that is foisted on the citizens from the official side. Neighborhood borders are not administrative borders, but have more to do with the subjective demarcations of the so-called "localities"[89] on the part of residents informed by individual experience and interpretation of the social-spatial environment. The neighborhoods are accordingly city areas in process and are geographically and sociologically an interim solution to a certain degree. City functions are increasingly delegated downwards in the American cities. This is especially true for social establishments, which the neighborhood organizations take over themselves depending on the degree of social interaction, community awareness and the access to external social institutions (Behrens 1995, p. 35). Content-wise, three dimensions can be differentiated: the physical-spatial, the social-behavioral, and the dimensions of cognitive identity. These parameters imply that the neighborhoods tend towards residential differentiation (Johnston 1978, p. 175). As motives for this, Behrens (1995, p. 51) identifies points out three fields: social welfare, political activity (to achieve a change in the neighborhood), and the preservation of status quo (a rather conservative opinion). Police, healthcare and garbage collection continue to be organized by the city, but the administrative borders seldom coincide with those of the neighborhood.

This specifically American organizational form has its historical roots in the early 20th century and can again be traced back to the maxims of American private capitalism. The organizations differ in social conditions, strategies, and tactics in practicing political influence. Different periods are represented throughout the history of neighborhood organizations. They go from so-called social welfare organizations in 1886 up to radical and conservative periods and revolutionary startups during the civil rights movement in the 1950s and 1960s (Fisher 1994). Generating conflict and building consensus are just as much a part of the methods of the different organizations as are the active intervention in the local development of the neighborhood and the support of an (internal or public) social plan. Impor-

---

[88] Also referred to as: Residents Association, Civic Association, and Neighborhood Association.

[89] Localities are multi-functional urban districts resulting from social networking. This gives to the inhabitants a unique social identity. Localities are amorphic districts such as a neighborhood.

tant in these organizations in the cities is the fact that inhabitants compete for space with other residents of the city.

The neighborhoods spread and develop at different times. Many neighborhood organizations were founded in the 1920s along with the City Beautiful Movement. A second wave of neighborhood organizations can be seen after the Second World War, as a new consciousness of one's immediate environment began, among other reasons due to the new and different stream of immigrants. A further wave began in 1960, when the situation of the cities became desperate and many cities were in the process of urban renewal (Urban Renewal Program) (Adams et al. 1991, p. 120). The neighborhood organizations offered the residents of a district city-planning participation and co-determination in decision-making as well as mutual support based solely on a shared living environment without open reference to religion and ethnic affiliation.[90] Many members also state that an important, fundamental aspect of active collaboration within a neighborhood organization is the preservation of quality-of-life issues. This refers to peace, quietness, civility and absence of crime, as well as aesthetic quality or outward appearance of the public and open space (Behrens 1995, pp. 91). Nevertheless, it is conspicuous that the organizations are found most often in wealthy areas with a well-educated population as well as in areas having a distinctive residential character, as is often the case with the historic districts (Cuttler 1980, p. 273). Following are classifications for the various neighborhood groups identified for the areas that are the subject of this study:

[90] More detailed studies reveal that this is, in practice, almost certainly not the case, though it remains a prominent feature of such organizations in theory.

Table 4.3. Types of neighborhood organizations in the four historic districts examined in this study

	Aesthetic difference	Social stabilization	Empowerment
Historic district	Society Hill (Philadelphia), Back Bay (Boston)	Savannah Historic District (outlying areas), Old and Historic Charleston District	Old and Historic Charleston District, (Outside the relevant historic districts, north of Center City Philadelphia)
Social Basis	Cultural elite, "Community liberated"	Middle class	Ghetto, lower class, "Community saved"
Housing market tendencies	Stable, gentrified	Instable	Affordable Housing Projects, tendency to become a slum
Priorities	Living comfort, municipal design (design guidelines), quality-of-life issues	Stabilization of housing market, promotion of house ownership	Redevelopment, social service
NIMBY	Strong	Varies	Weak
Dependency on the public sector	Low	Low	At times very high
Organizational form	Member association	Member association	Social support organization
Resident meetings	Rare	Regular	Regular

Source: Adapted from Behrens (1995, p. 143)

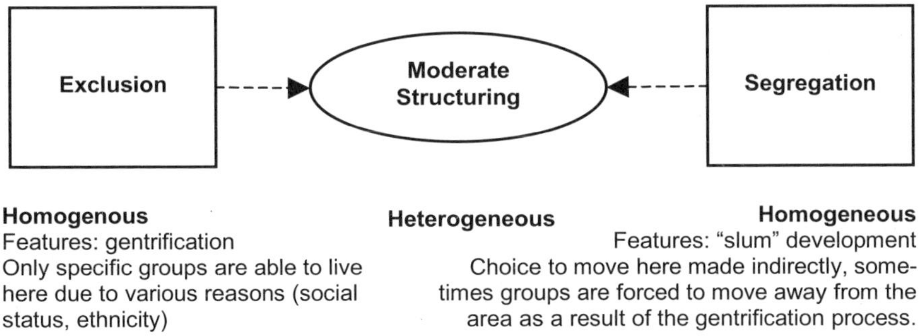

Figure 4.9. Exclusion and segregation

The neighborhoods' territorial controls are a process of the segregation of certain (unwelcome) groups (Weiler 1980, p. 229). A very active neighborhood organization is not necessarily interested in exclusive city patterns but can be a clear indicator of their existance.

An awareness of historic preservation developed in relatively few neighborhood organizations from the 1960s to the 1970s. Berry (1993) notes that the neighborhood organizations raise no new issues but are crucial in the municipal planning process (Berry 1993, pp. 157). Very active neighborhood organizations are found in urban areas in districts having an extensive, architectural historical legacy, including, for example, Society Hill, Philadelphia or Beacon Hill, Boston. A mature and long-standing community strengthens the practices and the influence of a neighborhood enduringly (Fisher 1994, pp. 30).

The influence and interests of homeowners in historic preservation varies considerably. Many neighborhoods have a tendency towards selective immigration, which means that only certain population groups move to these areas. This is also due to the politics of the neighborhood organizations, which an interested buyer often inquires about in advance.

Individual, achievement-oriented thinking and personal ambition can reinforce the motivation towards historic preservation activities. The quality of neighborhood contacts is another factor that determines the extent of historic preservation. Strong neighborhoods often anticipate the decisions of the public and political authorities. This is a part of the grassroots democracy.[91] Perception of others can also have a strong effect on the degree of historic preservation in a neighborhood.

Terretories that stand from surrounding neighborhoods enable their borders to be defined. This occurs extremely successfully in the context of the applicable laws of the historic districts. Active neighborhood organizations are found most often in the historic districts. The neighborhood organizations divide themselves into special areas called committees. The Beacon Hill Civic Association (BHCA), for example, is divided up as follows (BHCA 1975, p. 6):

1. Architecture
2. Street cleaning and air pollution
3. Ownership and tenant questions
4. Legal department/legal consultation/advice
5. Construction regulations and planning
6. Sidewalk and streetlights
7. Traffic and parking
8. Trees, plants, and general beautification

The active defense against disturbances and infringement on the urban quality of life is organized essentially through the strict monitoring of zoning ordinances. In the historic district, this extends to inspecting and enforcing the historic district ordinance or the historic zoning ordinance. To avoid disagreements and legal actions due to misunderstandings, the Beacon Hill Civic Association recommends that owners follow the procedures in Fig. 4.10 before starting construction. Of

[91] See glossary.

course, the subjective individual opinion of a neighbor or of a neighborhood organization is not legally binding on the owner and only the official local code or regulations are strictly applicable.

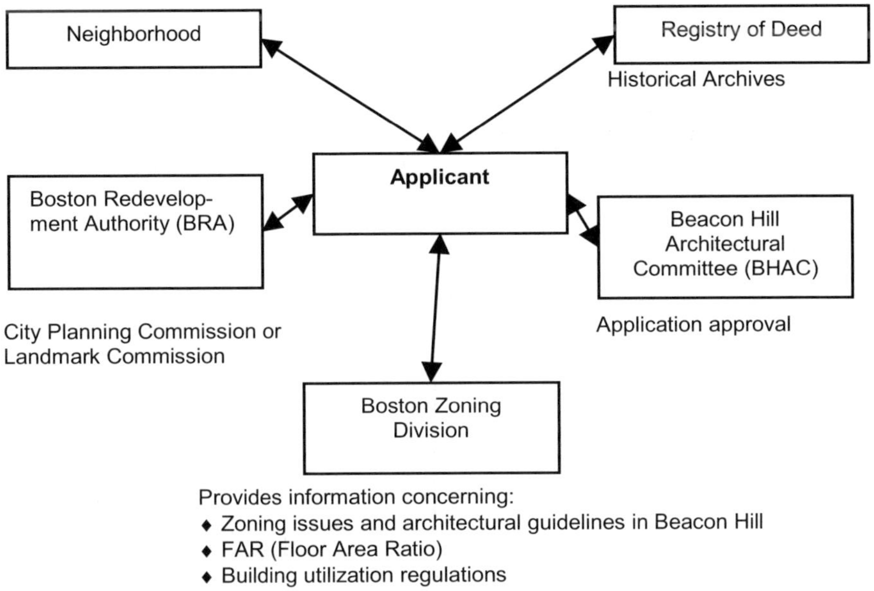

Figure 4.10. Recommended interaction for owners and tenants in Beacon Hill Historic District applying for building modifications (Source: based on BHCA 1975, pp. 5)

Figure 4.11 shows both the formal, legally prescribed procedure, together with the informal procedure that the owner is encouraged to follow prior to or parallel with his official application.

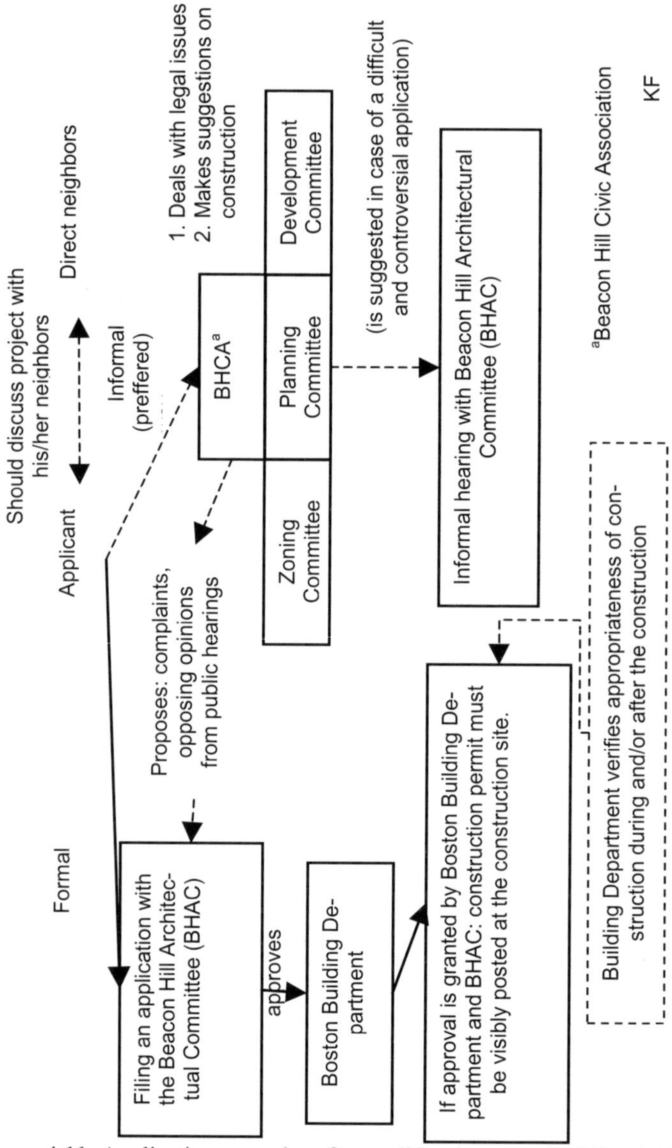

Figure 4.11. Application procedure for modifications to a building in Beacon Hill Historic District (schematic) (Source: based on BHCA 1975, pp. 5)

Some neighborhoods consider their area to be exclusively residential and make every effort to protect it against to other forms of use. This applies in particular to the "... basic rejection of granting new alcohol licenses in the gastronomic area..." (Society Hill Reporter 1999, pp. 4). Although Staubach (in: ILS 1999, p. 40) recognizes that neighborhood organizations are disproportionately active when it comes to obstruction, one must add that they do not join together and see themselves as opponents of the public sector in all instances.[92] This form of delimiting on the part of the private neighborhood organizations results in a conflict of interest in terms of tourist usage or revaluing historic neighborhoods. The residents of historic houses defend their own area vehemently and successfully against the onslaught of tourism in the historic district. In this case, the tourists represent a clear instability factor for the residents and are seen as a threat to quality of life. The neighborhoods do play an undisputedly key role in the context of successful preservation work. The accompanying massive gentrification processes, as well as the exclusive tendencies should not, however, be ignored.

4.5 Intermediate Findings: the Historic Districts in the American Urban System

4.5.1 Political-Administrative Evaluation Level

The degree of historic preservation in the American city depends very much on the local government. Historic preservation assumes the form of a very local and private matter in which the government provides little initiative, support or funding (see also Cullingworth 1997), p. 117). It is above all the local participants and micro-processes in the city districts and neighborhoods that push historic preservation forward.

Federal efforts in the field of historic preservation guaranteed no real protection before 1974. Owners of historic or significant houses were merely required to maintain their houses as they sawfit. Only on the local level is there greater discussion and influence regarding modification to and demolition of monuments. Within the historic districts, this protection is even more clearly ensured. Even new buildings in the historic district are visually adapted to the existing historic legacy, through choice of color, material or form.

---

92 Society Hill in Philadelphia also attempts to restrict liquor licenses.

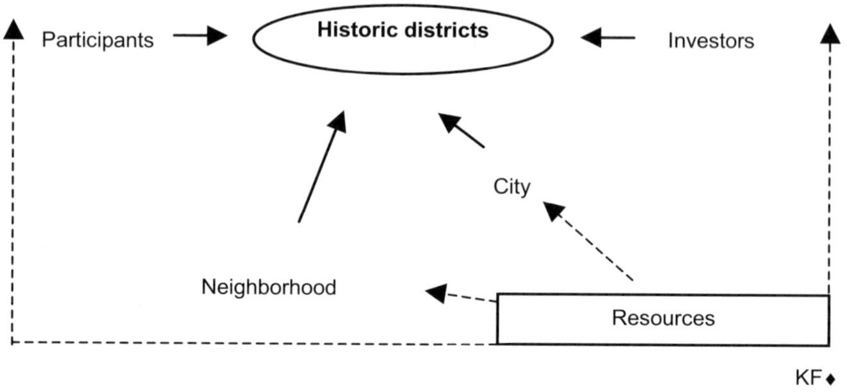

Figure 4.12. Processes in American urban districts

Nevertheless, monuments in the United States would surely not have been inventoried as systematically as they have been if the governments had not used legislation and state-level organs to intervene, and if the corresponding agencies had not been established. Given all the conflicts of interest, especially in the area of tourism, the neighborhood residential population and investors, governmental guidance has proven to be very helpful. Still, it became apparent that such guiding measures are strongly profit-oriented (as, for example, the listing and inventorying of historic structures) and are turned over to the individual state and local offices. Governmental and political restriction today prevent some premature demolition or structural modifications to historical structures, so that there is no guarantee of profitability.

It is also necessary to develop local historic zoning ordinances, together with the corresponding zoning ordinance adjustments, in order to take into account new developments and the interests of historic preservation. But participative planning on the part of the population in the sense of grassroots democracy is also promoted in the creation of these laws and guidelines.

A currently sensitive topic within the cities is the construction of new highways, as well as the sealing-off of existing highways in order to rejoin neighborhoods that had been cut off by road construction. For such purposes, private regional planning offices have been established in the SMSA of many large cities, for example the DVRPC - Delaware Valley Regional Planning Commission (1993), or the Chatham County Savannah Metropolitan Planning Authority (1969).

4.5.2 Economic Evaluation Level

It is debatable whether the successes of historic preservation are objectively measurable and whether it is possible to isolate economic effects. Scientifically, their consequences are difficult to prove, since there are so many different programs with various objectives for historic preservation in the core cities of America. Measuring success is therefore very complicated, which is not exactly helpful to the social acceptance and perceived importance of historic preservation in America. Traditionalists argue in favor of culture and history, developers emphasize mainly the economic successes. Frequently, these interests conflict, which is reflected particularly in public and private motivations (Cullingworth 1997, p. 113). This thesis of Cullingworth, however, applies only in a limited sense to the dynamic in the historic districts, since there the public and private interests are often identical, whereas private individuals attempt to withdraw from the sometimes very strict local laws (Iv-Phl-14) or the neighborhood organizations find their hard-earned quality of life issues threatened.

In 1976 the Federal government created a significant motivational factor for the population for promoting historic preservation. The Tax Reform Act of 1976 (public law 94-455) established important tax breaks for the preservation and rehabilitation of commercial and income-producing structures if they are recognized by the U.S. Department of the Interior as historic. In addition, however, penalties were established (for example, by increasing the tax rates) for the (knowing) destruction of historic structures. The tax write-off changed the social acceptance of historic preservation in the U.S.A. considerably, since now even private project developers could set up a profitable business with new plans incorporating historic preservation. Prior to that, tax law was so counterproductive that even the demolition of historic buildings could be written off. In such cases, reference was often made to the statutes of the Urban Renewal Program[93] (Cullingworth 1997, p. 129).

A year later (1977), as part of the National Neighborhood Policy Act (Title II of public law 94-24) an additional section was added to the tax concessions. This new paragraph stipulates that existing city neighborhoods are a national resource to be maintained and revitalized (Murtagh 1997, p. 213). This stipulation triggered another wave of incentive concerning historic districts at the end of the 1970s. Since then, various organizations have attempted to revitalize individual rundown neighborhoods in the cities with the help of these monies (Iv-Bos-3, Iv-Phl-11). Especially active in applying for these funds are the structurally strong, gentrified neighborhoods, on the one hand, and, on the other, those various public organizations that are particularly dedicated to promoting and supporting minorities in the cities. Both approaches will be covered and exemplified in the empirical section.

Another milestone in the area of tax concessions for historic preservation investments, and thereby for the indirect support of historic preservation, was with the Economic Recovery Tax Act, passed by Congress in 1981. This paragraph provides for a 25% tax reduction for the renovation of historical buildings. This law has had a significant impact on historic preservation, especially in urban areas. The Tax Reform Act, passed during the tenure of President Reagan in 1986, reduced these concessions considerably, but still did not discourage investors completely.

[93] The Urban Renewal Programs were initiated and managed by the local Redevelopment Authorities (RDA).

Today, every approved renovation and restoration project is encouraged through a 20% tax incentive[94]. State laws also support many investors and private individuals in historic preservation. All in all, there are six programs for supporting historic preservation (based on: Cullingworth 1997, p. 120):

1. Tax Reduction Program/Free of tax
2. Loan approval or interest reduction for renovations
3. Special calculation of property tax or tax rate
4. Income tax reduction
5. Sales tax reduction when selling the building[95]
6. Other tax reductions

The formulation and exact methods, however, vary considerably from state to state. Funds and donations from various private and business, as well as public agencies, which also receive tax concessions, are another common form of supporting historic preservation.

Tourism is a particularly significant factor with respect to the profitability of historic preservation. Bornheim (1956, p. 97) already emphasized the practical nature of museum culture in the United States. The most recent studies having a comparable component are described in the monograph by Schuck-Wersig and Wersig (1999). The conflicting goals with respect to the touristic industry will be presented and debated in the empirical part of the present study. Cullingworth (1997, p. 113) goes so far as to criticize the inclusion of American parks, which in his view are in danger of mutating into theme parks or zoos. Many local banks, convinced of the microconcepts of revitalization within the framework of historic preservation, make loans available for the implementation of individual projects.

Likewise, the Redevelopment Authorities (RDA), which were founded during the period of urban renewal (1950s and 1960s) and are still in existence in the cities, also have the ability to carry out historic preservation in some coties using local monies.[96]

Although many programs, possibilities and support projects exist for historic preservation, the main financial burden lies, as always, with the investors. Only financially better-off investors are able to use tax write-offs profitably and withstand even slight losses. As recently as the 1970s, historic preservation in America was described as follows: "Historic preservation is essentially a program that provides housing for middle- and upper-income people .." (Silver 1991, p. 80). This point of view can be revised today, thanks to a number of integrative projects. Fitch (1995, p. 78) also expresses the reservation that an excessively tourism-oriented invaluation of the historic districts can lead to the destruction of the ac-

[94] ITC: Investment Tax Credit.

[95] The sales tax for the following cities is as follows: Philadelphia 7%, Boston 6%, Charleston 6% and Savannah 6%.

[96] This was the case in Philadelphia, where an abandoned house in the Washington Square West district was renovated thanks to RDA funding (Iv-Phl-8). The analytical section of this study identifies similar cases in Savannah and Charleston.

tual historical structures, which in turn would stand in the way of lasting historic preservation.

As early as 1977, it was noted that, besides the legal basic questions concerning property rights, the main problem lies in the procurement of financial means for renovation (Hofmeister 1977, p. 146). This perception has not changed significantly in the meantime. But since awareness of historic preservation has heightened significantly, more funds have been made available, while the procedures for obtaining them have been simplified with the assistance of the many participants. How these economic components are profitably integrated regionally in the individual historic districts will be discussed in the following chapters.

4.5.3 Socio-cultural Evaluation Level

Our description of the participants in the field of historic preservation has shown that urban historic preservation is characterized by various social motives. Among the most influential participants for the historic districts are the neighborhood districts.

In summary, we note firstly that the historic preservation movement in America grew in disorganized fashion out of the grassroots democracy (Cullingworth 1997, pp. 116). The movement arose initially from individuals whose primary objective was to protect specific buildings. Many of these attempts failed because the financial means were lacking. A notable example from this era would be the John Hancock House in Boston, which was torn down in 1863. Independence Hall in Philadelphia was barely saved from this fate when the structure was sold in 1863 for the then ridiculously low price of $70,000. The National Parks, on the other hand, became part of public consciousness early on. An appreciation for architectural and historic preservation values in an urban context is not seen until the end of the 19th century. Prior to that, interest was limited rather to U.S. American history and the associated structures.

In 1978, the National Council for Preservation Education (NCPE) was founded in response to this imbalance. Among its activities, the NCPE awards grants for the encouragement of historic preservation including its goals, ideals, historic development and promotion, with an overall objective of promoting social awareness and acceptance. Educational materials have been created which can be used when needed in the curriculum of the individual states.

Historic preservation is, at the same time, a new source of enduring controversies. Historic preservation is driven by the local identity and selective moving habits of some Americans.[97] Different residents have varying expectations for the neighborhood in which they live. Some are interested only in fulfillment and as-

---

[97] A large proportion of the population looks for a place to live before they start to look for a job. Many who move house, for example, choose to settle in an historic district, because their quality of life will be improved (Iv-Phl-14). Thus, the historic districts can be seen to have a certain "hallmark of quality".

surance of quality-of-life issues; others identify themselves culturally with the neighborhood, which inevitably has historic preservation implications.
It should be mentioned again that the social value of historic districts is difficult to asses in monetary terms (Tiesdell et al. 1996, p. 27). This difficult question will not be addressed in detail here, but rather through the use of examples. The first look at a building is, in most cases, emotional, which is why historic preservation is a strongly cultural and emotional value, representing an essential part of heritage (Feilden 1997, p. 1). In this connection Feilden (1997, p. 1) differentiates between three values: the political, the spiritual and the symbolic.
Figure 4.13 summarizes the previously discussed influences in the historic districts.

Following the intermediate findings from the theoretical section, the assessment levels developed will be used in the empirical section as a basis for evaluating the four historic districts in the cities of Philadelphia, Boston, Charleston and Savannah.

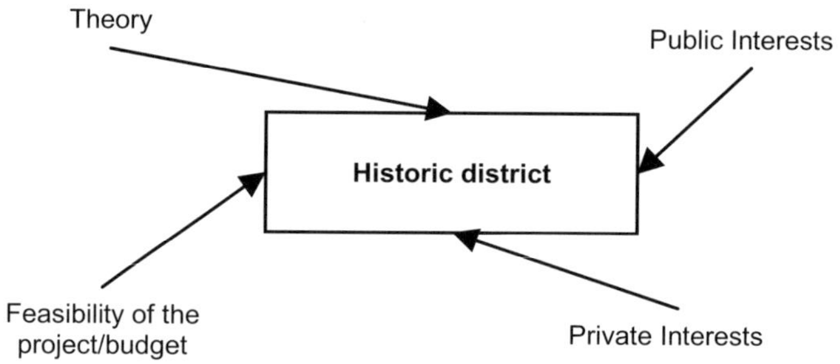

Figure 4.13. Influencing factors in the historic districts

5 Historic Preservation in Philadelphia

In the following empirical section, the historic district Society Hill[98] in the City of Philadelphia will be examined in detail. The relevance of the research in this study lies in the way in which historic preservation in Society Hill has been put into action and in the application of existing theories and methods. Philadelphia is appropriate for this case study because the city shows many of the structures typical of an old industrial metropolis within the manufacturing belt with declining or stagnating population numbers, and the historic district Society Hill produces an interesting dynamic within the city framework.[99]

---

[98] Society Hill is the name of the neighborhood. The district is actually called "Washington Square East". The correct name of the historic district is: Society Hill and Pennsylvania Hospital of Washington Square West. The hospital, although not located within the neighborhood of Society Hill, was originally established for the residents of Society Hill and was therefore incorporated into the name (Society Hill Reporter 5/1999, p. 3).

[99] Socio-historic background information can be found in Balack's historical analysis.

5.1 Preliminary Remarks and Maps of the Philadelphia Chapter

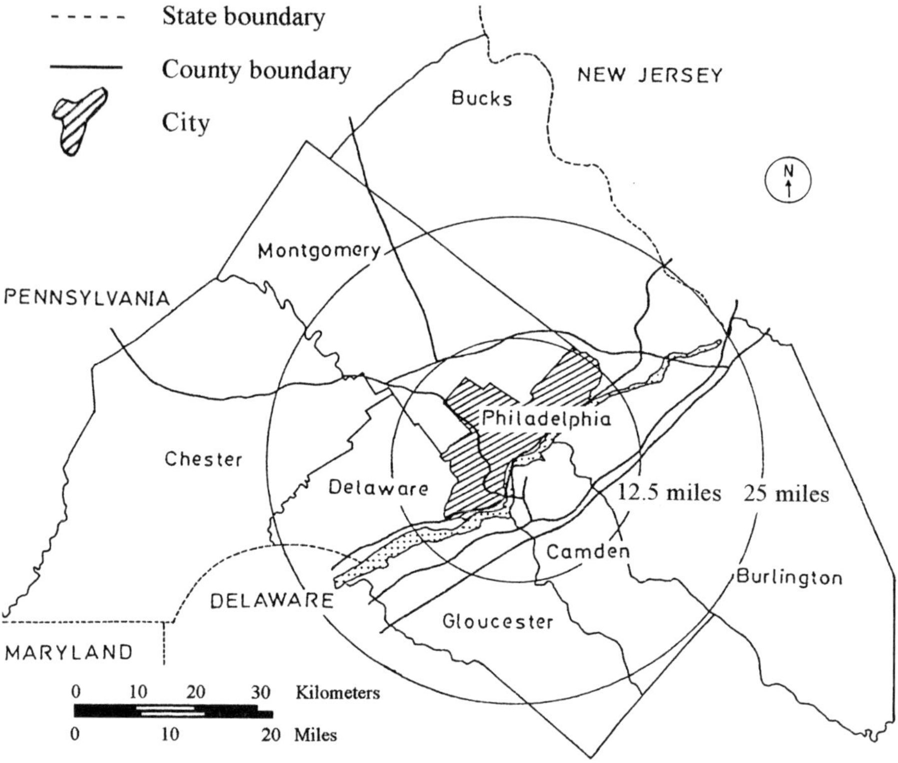

Map 5.1. Philadelphia SMSA (Source: Hahn 1990, p. 40)

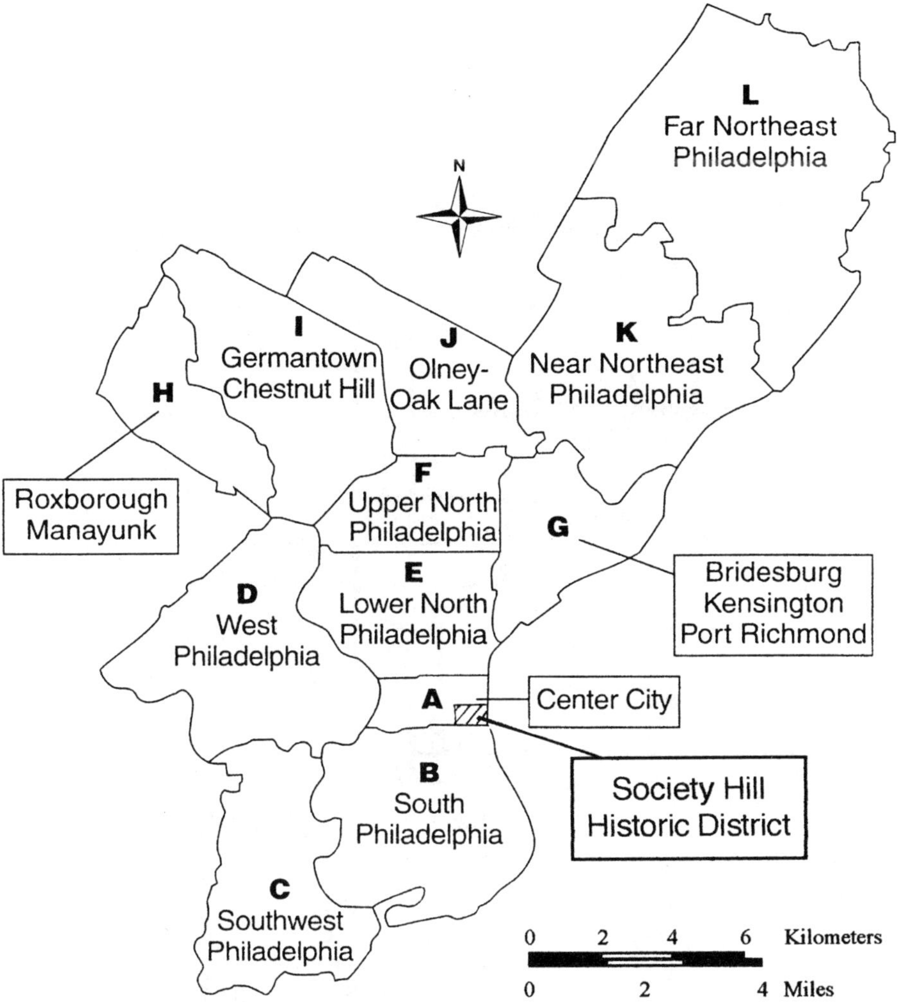

Map 5.2. Philadelphia City Planning Districts and location of Society Hill (Source: City of Philadelphia 1999, p. 28, modified by K. Frank)

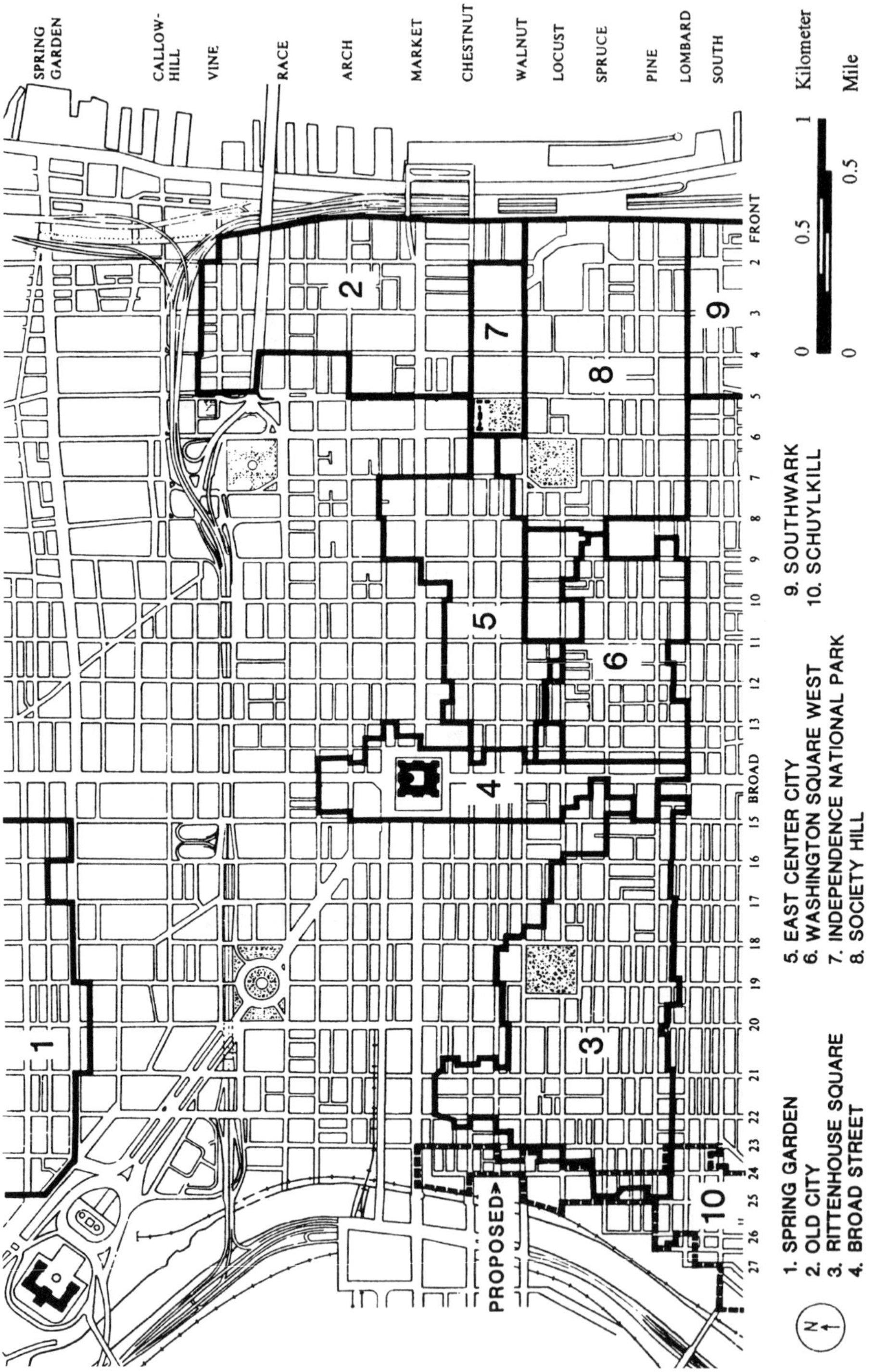

Map 5.3. Center City Philadelphia: Historic districts (Source: Phil Plan 1988, p. 27)

5.2 Urban Planning and Building Inventory

5.2.1 Urban Planning in Philadelphia

The names that have been given to Philadelphia throughout the years are a good indicator for its historical significance in U.S. history. These names include: The Quaker City, Cradle of Liberty, Athens of America, Workshop of the World, City of Homes, City of the First, Holy Experiment and City of Neighborhoods (Table 5.1).
William Penn, founder of the "City of Brotherly Love"[100], was forced to realize, even in his own time, that his "holy" experiment had failed. His religious brethren, the Quakers, had used freedom to behave just as egotistically and foolishly as had the supporters of the Anglican Church in the motherland England from which the Quakers themselves had fled.

Table 5.1. Naming for the City of Philadelphia

Year	Name	Meaning/Significance
Founding of the city (1662)	City of Brotherly Love	Translation of "Philadelphia" from the ancient Greek.
First settlement years (17th/18th century)	Quaker City	Many Quakers who were persecuted in England due to their religion came to Philadelphia.
War of Succession (mid-19th century)	Workshop of the World	The name goes back to many slaves who sought refuge in the city during the American Civil War.
Booming years (end of the 19th century)	City of the First	Institutions such as banks, department stores, zoos, hospitals and fire insurance companies were established in Philadelphia for the first time.
About 1920	City of Neighborhoods	The unity of Philadelphia which had been split by the diverse ethnic and social groups did function in the individual, essentially homogeneous neighborhoods.
1968	The Private City	Warner (1991) coined this name in his book: he portrays private capitalism in Philadelphia historically and thereby explains the so-called urban crisis.

The population of Philadelphia had, in comparison with other U.S. cities, initiated and supported a number of directed urban renewal and planning phases which are listed briefly in the following table.

---

[100] The name "Philadelphia" is a word-creation derived from the two Greek words philos=love and adelphos=brother/sibling.

Table 5.2. Names of city development and planning projects in Philadelphia

Year	Name	Comment
1910-1920	City Beautiful Movement	The uncontrolled nature of city planning was noticed; based upon Parisian architecture, this movement was initiated by the citizens, who placed the aesthetics of the city in the foreground. It was not a planned movement, but rather proposed only isolated planning projects (the Benjamin Franklin Parkway, for example, was modeled on the Champs Elysées).
1940s	Redevelopment Authority of Philadelphia (RDA)	An organization which was intended to control, and thus overcome, the beginnings of the city decline ("urban crisis") in Philadelphia.
1980s	Back-to-the-CityProgram	The rapid migration from the city in the 1960s and 1970s was supposed to be stopped or reduced through revaluation and revitalization measures.
1990s	New Urbanism	Directed towards further emphasizing the urban advantages of Philadelphia. Historic preservation is supposed to assist in expanding tourism and establishing a new, permanent residential population in Philadelphia. The historic districts play a central role in this.

Center City Philadelphia today features many open areas which the city administration would like to see used for economic purposes, but there is, on the other hand, a steady population loss which has continued unabated since 1950 (highest population level according to the U.S Department of Commerce 1960-1990).

The following shows a representation of data from the U.S. Census in map form. Breaking this data down further to the district level makes it clear that Center City, which also includes Society Hill, represents an exception (City of Philadelphia 1999, p. 28). A comparison with the same statistical records from the U.S. Census of 1970, 1980 and 1990 allows conclusions to be drawn regarding segregation and gentrification processes, spatial income level distribution and suburbanization tendencies, as well as the general economical development of Philadelphia. The strong structural socio-economic and social data for Society Hill are especially prominent in this portrait.

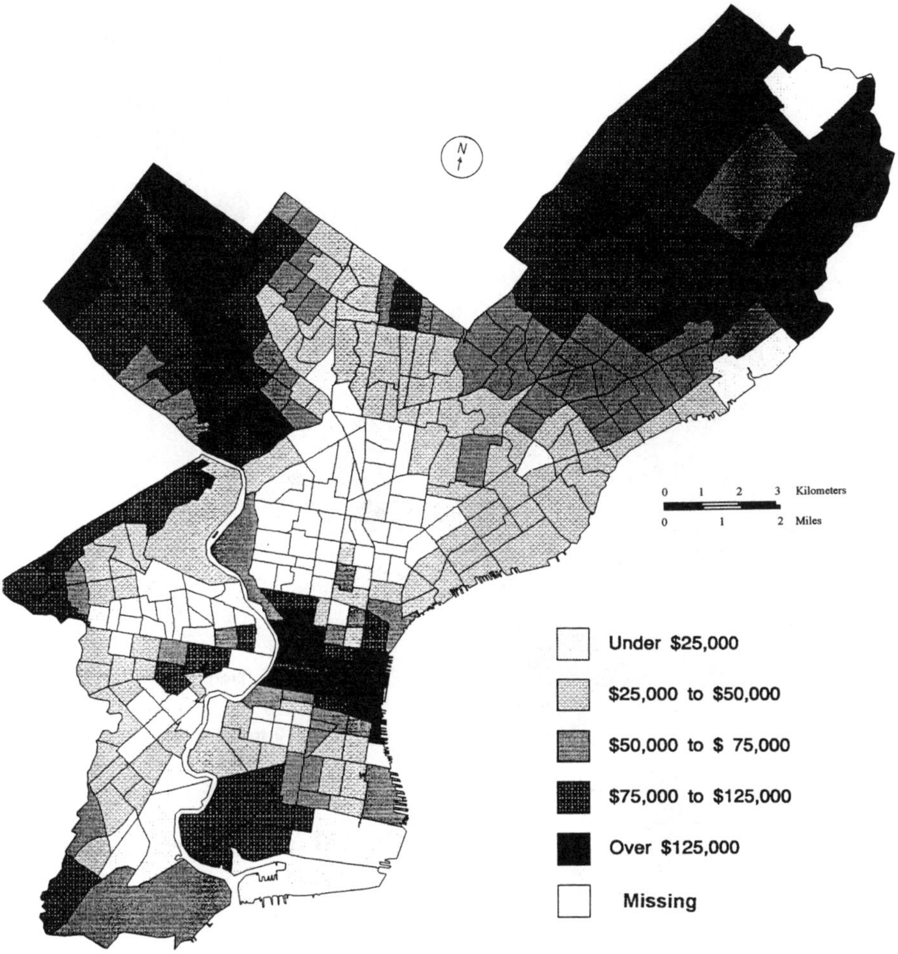

Map 5.4. Median housing values, City of Philadelphia (Source: U.S. Census 1990, modified by K. Frank)

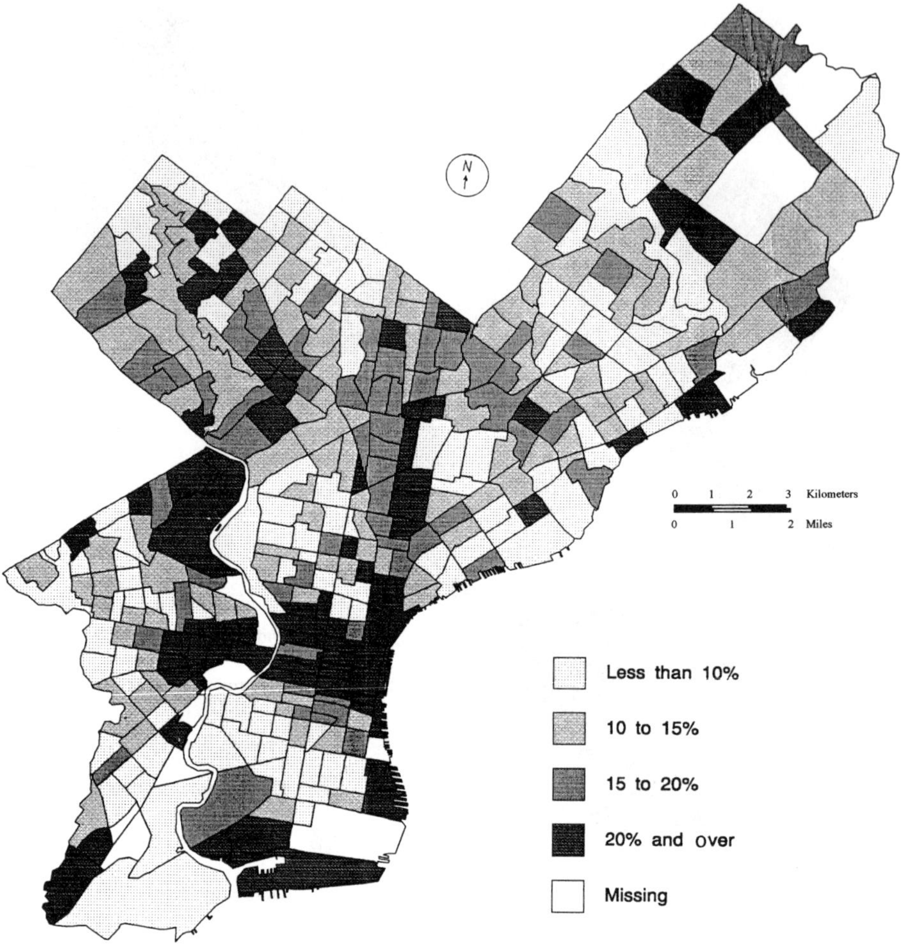

Map 5.5. Percent residents residing for less than a year, City of Philadelphia (Source: U.S. Census 1990, modified by K. Frank)

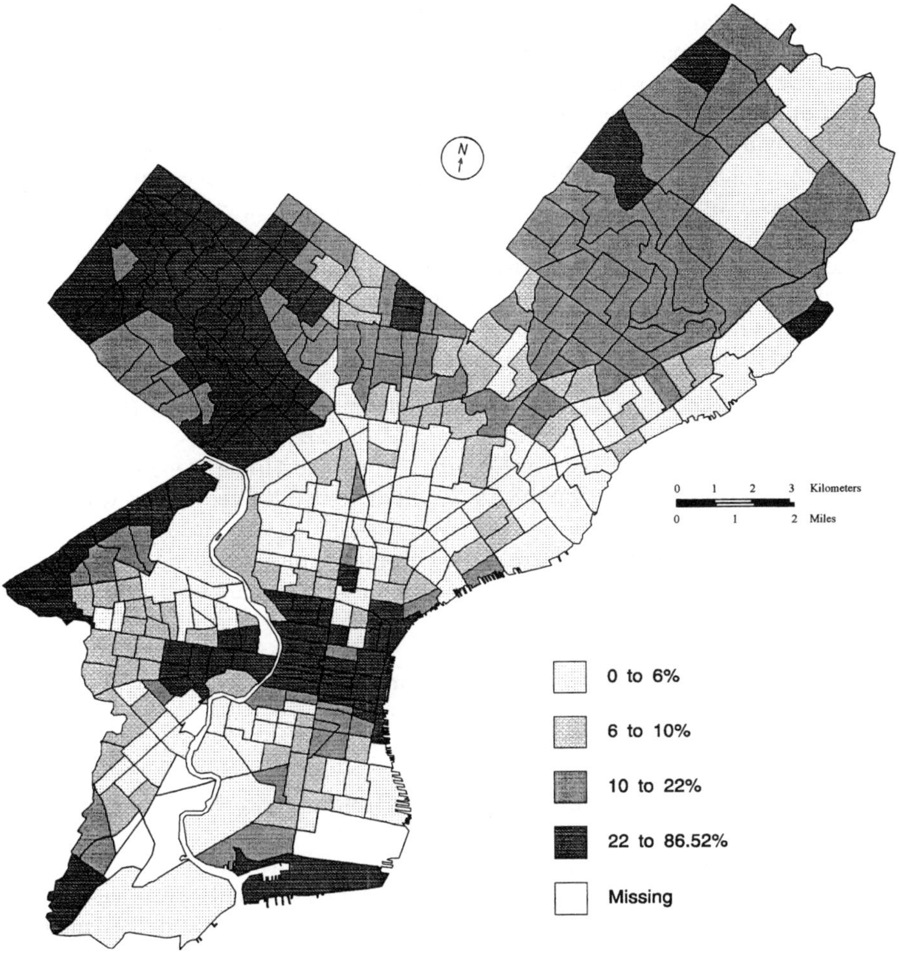

Map 5.6. Percent holding a college degree, City of Philadelphia (Source: U.S. Census 1990, modified by K. Frank)

Philadelphia is a city in which the traditions and maxims of private capitalism clearly prevail. Warner examined the general origins of the city of Philadelphia with respect to private capitalism (Warner 1991). His main thesis is that private capitalism, with its laissez-faire policy, was mainly responsible for the decline of Philadelphia. Factories and industries which were no longer profitable were quickly shut down, which in turn took capital away from that area (see also: Smith 1979). The resulting unexploited industrial landscape was subject neither to a collective responsibility nor to a governmentally directed city planning concept. There was no higher planning commission to financially or legally control the redevelopment, rebuilding or demolition of disused factories or abandoned and run-down houses. Since the public budget share depends directly on the number and profitability of entrepreneurs in each respective city district, redevelopment which did occur in certain areas was self-regulative, depending on need and budget. If this share is very small in structurally weak areas, it is difficult, if not impossible, to make or implement plans. In the financially strong Center City areas, many of the old buildings have been torn down and some turned into parking lots or carparks. A small area is being reused as commercial land. In structurally weak areas, as, for example, in North Philadelphia, demolition took place only when the building was in imminent danger of collapsing and public safety was at risk.[101]

Historic preservation in Philadelphia has already been incorporated in different city development phases as an (urban planning) instrument. This occurred above all with the objective of boosting the city (Table 5.1). The micro-phases, which at various times have helped to boost revitalization and historic preservation in Society Hill and Philadelphia since 1950, are as follows:

1. 1950 Urban Renewal Programs by the Redevelopment Authority (RDA 1968, 1980).
2. 1963 renewal/redevelopment of Society Hill by the Urban Renewal Program, resulting in gentrification (Smith 1979, 1986). A "spill-over" effect was hoped for in the surrounding areas.
3. In 1990, the Philadelphia City Planning Commission introduced the "Plan for Center City" (Phil Plan, 1988a), in which the nearly-bankrupt city was supposed to be revitalized economically and socially.[102]
4. Currently: City establishes empowerment zones and various historic districts. This also takes place in structurally weak areas with less strict or optional legal protection conditions.

5.2.2 Continued Existence of Historical Buildings in Philadelphia

The historical tableau of Philadelphia is very broad and is characterized by many significant buildings from the 17th to the 20th centuries. Virtually every era of U.S. American architecture is represented in the architectural variety of Philadelphia (McCall 1966, see also figure in Appendix A.3). Many well-known architects have left their mark on Philadelphia, including Pei, Khan and Scully, to mention just a few. Philadelphia has always prided itself on having more historical buildings than any other city. On closer analysis, this initial claim is seen to be quite subjective, especially when one considers that there are no nationally recognized standards for defining at what point a house can be considered historic. Still, it cannot be denied that the entire city area of Philadelphia does contain a considerable number of historic houses. In most cases, however, the buildings have not received historic protection status or been statistically included in the Philadelphia register (Iv-Phl-11). Only in areas with an active and committed neighborhood or-

[101] The daily press reports regularly on how dilapidated houses collapse and cause injuries following heavy rainstorms, for example (Philadelphia Inquirer 22ond of October, 1996).

[102] Frank (1997) carried out a detailed study based on four major projects in Philadelphia which were part of the "Plan for Center City".

ganization (spatially limited), have complete inventories been drawn up.[103] Philadelphia also contains a considerable number of churches and other sacred buildings. Various communities have for some years been working together on special historic preservation programs.[104]

The many paleo-industrial buildings in Philadelphia first convey a nostalgic impression which recalls the old days of a flourishing economy. The old warehouses and manufacturing buildings, on which faded inscriptions can often still be read, actually represent a great burden for Philadelphia. The uncontrollable crime that takes place in these buildings, as well as their dilapidated conditions, are not a positive distinction for these buildings. Demolition is very costly, however, and often the ownership rights are not clearly defined, or the owner resides in another state and has no interest in the property, which, in the meantime, has become worthless.

Problematic in such cases is the question of new utilization. Philadelphia has more vacant buildings than potential new utilizations or investors for such costly and economically risky plans. Several old industrial plants have been redeveloped into publishing houses, fitness centers, carparks or offices. One common policy is new utilization as apartments.[105] Loft apartments in the Old City have become very desirable within recent years (Tiesdell et al. 1996, p. 115). New utilization plans often stumble over difficult problems such as utility access.

Nor is indirect historic preservation, in the form of tax write-offs, associated with historic preservation laws, or the incentive for new investors to locate a new firm in a nostalgic building, a sufficiently great attraction. The preservation of the building itself is evaluated only on the basis of economic profit and usually dismissed as unprofitable and of secondary importance. Further laws and regulations obstruct this situation considerably: legislation in Philadelphia provides that sufficient parking spaces be provided. This often involves a considerable intrusion into the building itself. The lower floors of several buildings have been gutted, so that the historical structure is no longer recognizable to passing pedestrians (see AIA 1965). It is a frequent source of regret that these regulations have led to a situation where pedestrians often see only automobile grills, and not the historical facades of buildings (Iv-Phl-9).

Yet, all too often, historical houses are hastily demolished due to economic obstacles, or are not restored according to the regulations. The authorities frequently have to remind owners of their historic preservation responsibilities in order to reduce the number of transgressions and take action when necessary to insure historic preservation (Uhlig 1977; Iv-Phl-16).

---

[103] The long tradition of having so many neighborhood organizations is also unique to Philadelphia. These private forms of self-administration, which are entirely unknown in Europe, are mobilized in mutual cooperation with the city.

[104] Particularly involved in this area as a national organization are the Partners for Sacred Places (1998), since sacred buildings are excluded from the federal historic preservation programs. This is a consequence of the strict separation of church and state in the United States.

[105] In South Philadelphia, an unused synagogue was converted into apartments.

5.3 Philadelphia Historical Commission (PHC)

Philadelphia has its own historic preservation office, the Philadelphia Historical Commission (PHC). The PHC was founded in 1955, and today has seven employees. It is an independent department with no administrative, financial, or technical connection with the Philadelphia City Planning Commission. Dr. Margaret Tinkcom was the first director of the Philadelphia Historical Commission. Under her supervision, all existing archive material for the registered historical buildings of Philadelphia was thoroughly sorted and organized. She worked closely together with HABS (Webster 1981) and expanded the national list to include many local buildings and historical city artifacts.[106] There are about 9,500 buildings in Philadelphia, 4,000 of which are to be found in Rittenhouse Square and in Society Hill, which are registered on the current list (Iv-Phl-11).[107] Insurance papers and countless notes written on the contracts of the houses from the 18th, 19th and 20th century by contractors and architects make Philadelphia's archive and inventories quite complete, though they may initially appear disorganized.[108]

The national registration of buildings in Philadelphia is restrained by comparison: From 60 original buildings dating mainly from the 18th century the HABS list for Philadelphia grew to include 600 by 1959. In addition the city was divided into nine historic-geographical districts.

The Philadelphia Historical Commission also administers the public city archives for all historically declared buildings. Furthermore, it informs owners about the current regulations which need to be complied with when renovating historical buildings and recommends additional historic houses for inclusion in the local historic catalogue. The files for individual houses consist of all the documents that exist pertaining to that building. This includes, among other things, an archive of plans, purchase contracts, photographs and all relevant correspondence. All four modification categories (rebuilding, tearing down, demolition or renovation) have to be submitted in the form of a written application to the PHC (PHC Br., archive material 1997). The Philadelphia Register also includes applications for modifications, records of justifications and exceptional conditions in the case of rejections and approvals. The social acceptance of historic preservation can be documented statistically as well: in 1997 there were 750 applications, compared with 993 in 1998.[109]

With respect to source material, Philadelphia is rich in primary sources: The first fire insurance company in America, which was founded in 1752 in Philadelphia, wrote an inventory report including the year of erection and condition of

[106] Especially noteworthy here is the wrought iron work and the striking Benjamin Franklin Light used for street illumination. This list also includes sacred buildings associated with various religious groups.

[107] "Only" around 12,000 objects (including historic districts) are listed in the National Register.

[108] Tinkcom especially studied strongly segregated and socially demarcated areas in North Philadelphia, such as the areas of Northern Liberties and Kensington.

[109] Conversation with Mrs. Spina (in: Iv-Phl-11).

every house insured against fire. This means that, in addition, the contractors and/or first owners of historical houses are in most cases known. Whenever a building has been classified historically valuable, the owner is informed. He can file an objection to the state proposal within 30 days (Iv-Phl-16). This is frequently the case if the owner feels financially disadvantaged by this proposal.

National as well as local historic preservation refers only to the outward apperance of the building in question as seen from the street (PHC 1997, pp. 27).[110] Interior architectural work is only seldom subject to historic preservation (ibid. p. 23).[111] Historic preservation works only with the existing physiognomy of the buildings, meaning that owners are not obliged to perform active architectural work to conform to a certain era, for example (Iv-Phl-11).

Historic preservation and the associated legal constraints only become relevant when the owner decides to undertake renovation work or make modifications to the buildings apperance. These have to be explained in written form and must be approved by the Historical Commission. The restoration must conform with the standards of historic preservation (City of Philadelphia 1984, Bill No. 318, App. No. 566, p. 15).[112]

PHC decisions are problematic and arbitrary in cases where the owner decides to do repair work on a facade which is non-historical but already existed before the listing. Occasionally this occurs when, for example, the air conditioning or outside steps are repaired. The subjective decision as to whether this non-historical part of the building has to be repaired or must be replaced by historically consistent structures is a matter for the PHC (PHC 1984, p. 50).

Pennsylvania placed (at first)[113] a legal prohibition on unilaterally declaring any of its cities as historical districts, which did not make the situation of historic preservation any easier for Philadelphia. The Philadelphia Register offered a solution by requiring a permit at least for the demolition of buildings (Gutschow 1978, p. 139). Ever since then, there has been a certain measure of directed control for historic individual buildings. The main task of the PHC consists of nominating, determining and identifying those structures of historical value which are worth preserving (PHC 1997, p. 5). The PHC also manages the Philadelphia Register, where buildings in the National Register and local nominations by the PHC are entered together (City of Philadelphia 1984).

Most of the nominating initiatives are generated by the population, typically representing individual neighborhoods and often with the goal of establishing an historic district. Many of the scientific inventory rules and design regulations have been developed by the neighborhoods themselves (Preservation Alliance 1998a, Iv-Phl-14).[114] Subsequently the PHC prepares them for binding legislation applying to everyone in the historic district which has come under protection. All

---

[110] This includes: Painting of the building or the fasade, windows, fences, etc.

[111] For example, the historic houses operating as museums.

[112] The only exceptions today to historic preservation are the fire protection measures.

[113] This provision no longer exists, but the intention remains the same.

[114] Agreements are currently being signed in for the Society Hill Historic District.

changes to protected houses which are visible from the street require a permit according to a fixed procedure administered and organized by the PHC.[115]

An interesting undertaking initiated in 1998 by the PHC is the Historic Street Designation Program.[116] In October 1998, the PHC initiated Section 14-2007 of the Philadelphia Code for the historical nomination of the various, still-existing street pavements (Historic Street Paving Thematic District). Numerous Philadelphia streets were incorporated into this initiative, streets whose paving has an historical character, with 320 blocks and a total of about 35 neighborhoods participating in the project.[117] Many neighborhood residents were surprised to learn of their nomination. This has contributed to an increase in the self-confidence of the residents in demographically weak districts in Philadelphia. For the first time, these residents feel that they are part of the city (Iv-Phl-11).

For the purpose of the nomination, every resident of the streets located in the district were invited to a public hearing, which is required for any nomination. The nomination procedures are similar to those for the historic districts, although the financing of the street paving is clearly a city matter. That also means that once the nomination has been approved, the city is obliged to present the proposed repaving to the PHC and construct it according to the historical model.

5.4 Historic Preservation Programs in Center City

Shortly after the "opening up" of Independence Hall, which involved tearing down several houses that were located on the prestigious forecourt, Society Hill was implemented as the first major district redevelopment. Several buildings were demolished and new constructions were added to the old structure. The intent was for the city to take a clear stand and to stem the massive flight to the suburbs (Iv-Phl-4). Even though it was not explicitly a matter of historic preservation, it did serve the purpose of recognizing the value of the historical edifices and in part restoring them. Smith (1979) confirms this process for Society Hill. He stresses (ibid. pp. 539) that the residents of Society Hill have not so much retreated from the suburbs as moved to Society Hill for other motives. He cites above all younger professionals who made a decision not to flee to the suburbs (ibid. p. 540). Cybriwsky (1986, p. 29) contradicts this argument by writing that these initiatives have brought back the first families from the suburbs.

Society Hill is characterized by houses from the colonial period and the 18th and 19th century (Fuqua 1997, p. 3 and p. 11; Photograph 5.1). In addition to these historic districts revalued by the city especially for tourists, there are other historical districts in Center City Philadelphia which are subject to other historic preservation procedures and dynamics. This applies, for example, to the neighborhood of Rittenhouse Square, which, even without nomination as an historic district, has,

[115] This administrative procedure is similar in all U.S. cities.

[116] Tunnard (1975, p. 90) has already addressed this topic.

[117] Conversation with Mrs. Spina (in: Iv-Phl-11).

with the help of the Preservation Alliance for Greater Philadelphia, already published an extensive brochure focusing on styles and restoration measures (Preservation Alliance 1998a).

After three years of planning and a number of targeted preliminary publications, the Philadelphia City Planning Commission in 1988 finally submitted final plan for Center City (Phil Plan 1988a). This plan goes far beyond the economic interests of Center City. It considers the historical past as well as present and future-oriented planning.[118] In this context, the city of Philadelphia submitted a further study in 1999, in which the economical and planning future of Philadelphia had been thought out and linked to the results of the 1988 plan (City of Philadelphia 1999). Above all, historic preservation, recreational areas, and urban renewal are emphasized in the new plan. Accordingly, the physiognomic concerns of Philadelphia no longer rely one-dimensionally on private initiatives (Phil Plan 1986a, 1987a, 1990c). It is thus clear that state control is becoming increasingly recognized.

The new plan, entitled "New Urban Direction", continues to see historic preservation as an important way of strengthening the internal and external image of the city. Historic preservation is carried out to a greater or lesser extent, especially in Center City, the Central Business District (CBD). Here we find the so-called "Historic Mile", in the neighborhood of Society Hill and Independence National Park. Society Hill has been designated a Local Historic District controlled by the PHC since March 1999, whereas Independence National Park is under the National Park Service.

The Financing and implementation of the individually proposed projects in the plan are intended to be the responsibility of mainly public-private joint ventures. Therefore, organizations have been founded specifically for each individual project.[119] These are undertakings aiming to make a profit, which operate according to private capitalist principles, and are expected, depending on their own defined range of activities, to assume responsibility for revitalizing, buying and selling of property, and advising investors. In order to counter the risk of an unintended laissez-faire tendency and to ensure that everything is implemented in accordance with the city planning office, the latter is given final power of disposition.

These different forms, motivations, investments, and revaluations of historic preservation in Center City are based on a variety of factors, which are examined in the following, using Society Hill as an example.

[118] Frank (1997) carried out an assessment of the success of this plan using four empirical examples in Center City Philadelphia.

[119] The largest in Philadelphia are: CCD (Center City District) and PIDC (Philadelphia Industrial Development Corporation).

5.5 Society Hill: Local Historic District

5.5.1 History of Society Hill

Society Hill is located in the southeast part of Center City and is among the oldest neighborhoods of Philadelphia. Located within an area of approximately 25 blocks are around 700 buildings, 500 of which are in the historical colonial and early post-colonial style (Garvin 1996, p. 215).

Initially, Society Hill was populated by the political and social elite, the members of the Free Society of Traders of Philadelphia (Weigley 1982). The maritime trade made the neighborhood wealthy, which is obvious even today in the imposing and beautiful houses to be found there. After the flourishing of the neighborhood, during the three city development phases of Philadelphia in the 18th and 19th century, the quality of life in that district declined rapidly. The well-off population moved to the newly created suburbs, initiating the decline of the previously exquisite residential area. Shops, workshops and warehouses were added, spacious homes were converted into commercial buildings, apartment houses, or even into mere sleeping lodges, which resulted at first in a mixed use area. The square on Dock Street in the middle of the quarter was used as wholesale market, bringing with it considerable traffic, noise and foul air. The once prime residential area of Society Hill was considered "dirty" by the mid-19th century (Cybriwsky and Western 1982, p. 356). Private investments made it possible, in 1930 and 1940, to maintain the area in its original form and to keep most parts of it from demolition (Tunnard 1975). In the absence of credit grants, liquidity, and ultimately lacking the interest of the population, those measures proved to be insufficient (Uhlig 1977, p. 43).

5.5.2 The Society Hill Project

Society Hill in particular was designed where and in the way it was precisely because the planners were against the decentralization of Center City (Baumann in: Krueckeberg 1997, pp. 182). As a result, the planning awareness for "living in the city" came into being (ibid. p. 187).

In the following years, several plans were presented which proposed to upgrade the dilapidated part of the city (NDP – Neighborhood Development Plans) (Garvin 1996, p. 214). Such plans required state assistance in order to make the neighborhood livable: It lacked schools, parks, parking and shopping facilities. The Redevelopment Authority of Philadelphia (RDA 1969) ultimately functioned as a patron for the renewal and upgrading of Washington Square East.[120] Society Hill was declared a National Historic District as early as 1955. The objective of the state was to preserve the increasingly rare historical roots of the U.S.A. Since Society Hill is not a discrete city district, a demarcation especially for the purpose of list-

[120] It is precisely such complete planning which makes Society Hill so unique in American history (Fitch 1995, p. 37).

ing it in the National Register was devised. At that time, the federal government was building up a countrywide network by publishing the first information, guidelines and clarifications concerning various architectural styles. At the same time, many buildings were added to the HABS list. Some buildings were saved from demolition due to the listing acknowledging their historical value (Webster 1981, p. 37). Nevertheless, from the classical viewpoint of historic preservation, this could not be considered effective protection. There was hardly any historical or architectural information available about the district, nor was there any means of funding or project ideas to protect the neighborhood completely.

Besides the image enhancement which went along with recognition as an historical district, Society Hill did not receive any direct financial means for historical conservation work. The 20 % tax reduction referred therefore to all redevelopment and preservation work within the district, independent of historic preservation. The federal government left overseeing of the historic preservation extent of the redevelopment to the city, in this case the RDA and the Historical Commission. By 1965, Society Hill was still a community in the making (Iv-Phl-1, Iv-Phl-2).

The physical renewal of Society Hill began in 1965. As well as the many observers who voiced their skepticism concerning this proposal, it was also very difficult to obtain bank loans for repairing the houses. At first, there was no state support at all for the restoration of authentic historic facades.

Dock Street Market, once the busiest location in the quarter, was relocated to the south of Philadelphia as part of the RDA's first initiatives. The idea was to keep the daily commuter traffic, the market visitors and the noise away from the newly created neighborhood (Garvin 1996, pp. 214). Located on the former site of the market are the three grand Society Hill Towers with apartments, designed by the famous architect Pei, and 24 additional newly built row houses, which were intended to symbolize the new Philadelphia (Photograph 5.2).

Critics attacked the building of high rise structures, saying they would spoil the skyline of Society Hill. Apart from the aesthetic argument, the new buildings offered so much residential space that the influx of people caused a rapid social change in Society Hill. This in turn had a positive effect on the image of the newly created neighborhood and attracted even more home buyers and investors. By 1972, just 16 years after this major project was begun, a total of 700 buildings had been renovated (Tunnard 1975, p. 89), some of them by former or new owners and others under trusteeship through the RDA, which resold them after the redevelopment measures (revolving funds).

Prior to the renewal measures in Society Hill, the population was an ethnically and, to some extent, socially heterogeneous mix. According to Cybriwsky (1986, p. 28), around 7,000 people lived in the district in 1940, most of them illegally and in utter poverty. Of these, approximately 20–25 % were African Americans. The rest consisted of European immigrants or members of ethnic minorities.[121] This great distribution was unique even for those times, when segregation of the city districts was just beginning (Goode and Schneider 1994).

[121] The Civil Rights Act of 1964/65 was not yet in force at this time.

Table 5.3. Selected demographic data for Society Hill in 1950 and 1980 (Source: Cybriwsky 1986, p. 29)

Demographic feature	**Society Hill 1950**	**Percentage City of Philadelphia 1950**	**Society Hill 1980**	**Percentage City of Philadelphia 1980**
Population	6,982	0.3% of the total city	5,213	0.3% of the total city
Average family income	18,600 US $	62.9% of what an average Philadelphia family earned	41,412 US $	Increase of: 252.7%
Percentage with high school degree	23.9%	--	91.8%	--
Average home value	5,224 US $	--	168,300 US $	Increase of: 716.2%
Percentage of houses without running water	24.1%	--	0.7%	--

Society Hill also had a large proportion of older inhabitants (Cybriwsky et al. 1986, p. 30). Reflecting this heterogeneity, the area was characterized by a large number of businesses and important cultural and religious institutions which strongly defined community life. Table 5.3 shows how strikingly the demographic data for Society Hill changed compared with Philadelphia overall between the pre-renewal date 1950 and 1980.

It was to be foreseen at the outset of the process that financial considerations would prevent some inhabitants from remaining in the area. Homeowners were paid $15,000 more than the current assessed value, and many tenants were forced either to pay drastically higher rents or to move out. Special efforts were made, however, to retain some residents, especially older whites.[122] They were offered free or subsidized housing paid for by public funds or donations.

There still does not exist an objectively written document issued by a higher authority concerning the exact details of buying and selling those homes or the other measures taken.[123] According to official figures, 483 families and 551 individual persons were relocated (Garvin 1996, p. 216). Where the inhabitants who were driven out moved to or how the social fabric in Society Hill has changed has not yet been examined in detail. At the time, there was probably no clear awareness of the extent and the consequences of the displacement process, especially since it concerned ethnic minorities who did not yet enjoy equality (Garvin 1996, p. 216). Many probably moved a few blocks further south to South Philadelphia or north to Old City and North Philadelphia. Some lost their means of living because

[122] This was intended to prevent a complete population reversal from taking place in Society Hill within a few years (Cybriwsky et al. 1986, pp. 27).

[123] Several unfortunate individual cases were reported to me orally.

they were either owners or employees of one of the numerous shops (Iv-Phl-2, Iv-Phl-5).

Birch and Roby (1984, p. 198) describe Society Hill as a brilliant testimony to the new appreciation for the common interests shared by historic preservationists and city planners (ibid. p. 198). This opinion is probably overstated, since Society Hill was not planned explicitly based on historic preservation considerations but rather was initially subject to the Urban Renewal Program. Society Hill was functional because it became a homogeneous part of the city following the major RDA project (Birch and Roby 1984, p. 199). Particularly in the 1960s this was a significant factor which led to success: It was considered chic to live in Society Hill (Gutschow 1978, p. 101).

The Society Hill project has been and continues to be treated in literature and the public as one of the most spectacular examples of the Urban Renewal Program (Tunnard 1975, pp. 8 and Uhlig 1977, pp. 42). It is not uncommon to hear about a house in Spruce Street which was sold for $3,000 in 1958 and today has a value of $1 million (Cybriwsky 1992, p. 30). Today Society Hill is one of the richest neighborhoods in the Philadelphia SMSA, with the highest average income, the highest house and property values, and an enormous tax base which benefits the whole city.

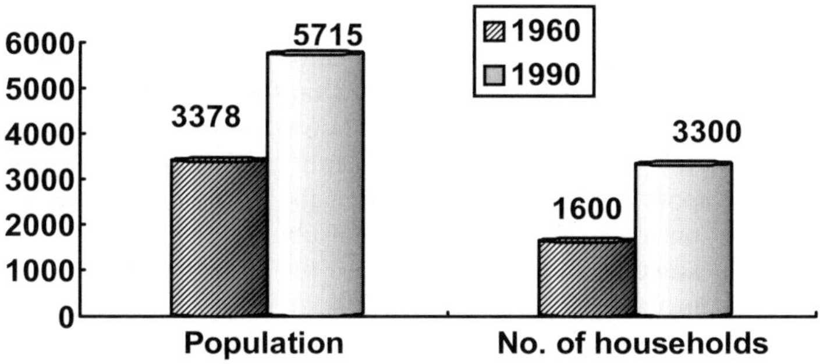

Figure 5.1. Population and number of households paying tax in Society Hill, 1960 and 1990 (Source: Garvin 1996, p. 216)

5.5.3 Sample Cases: Historic Preservation as Part of Urban Renewal

5.5.3.1 130–132 Delancey Street

As the first example of historic preservation work in Society Hill, we shall consider the historic colonial style house at Delancey Street 130–132. The building was constructed between 1756 and 1761, and has changed hands (renters and owners) 23 times since then. This list includes sailors, haberdashers, ship lookouts and retailers. In January 1958, the RDA purchased the house from a widow named Jennie Field for $1,100. The house was evidently in poor condition, so that considerable restoration and redevelopment work was necessary. The RDA had drawn up a plan in which the necessary modifications were listed in great detail for restoring the residential house to current, adequate living standards without destroying the historical facade at the same time. The interior architectural changes were only to be understood as recommendations for the buyer. It was left to the future owner as to whether the house was to be gutted or renovated. The renewal of the door and its frame, the window frames as well as the cleaning and reconstruction of the facade with its typical red mural stones were carried out as part of this process. The colors used to repaint the restored wood and metal on the outside wall had to be approved by the RDA.

Overall the list prescribes that the reclamation be done in a similar style to that of the adjacent house in order to maintain a uniform street appearance. The yard belonging to the house (Delaney Street 132) was to be encircled by a wall whose color and material had to be compatible with the facade. The yard landscaping also had to be approved by the RDA. Before the future owner was allowed to begin the restoration work, all plans had to be submitted to the RDA. The house remained legally however in the possession of the city of Philadelphia until the redevelopment approval was given by the RDA.

The present appearance of the house dates from two restorations completed by the owners in 1987 and 1988. The necessary building permits were granted by the Historical Commission after two and three weeks respectively. The iron door, the window shutters, and the outside lighting of the house were restored in 1987/88. The Historical Commission however found fault with the final restoration work because the previously approved plans were not consistent with the final facade: the owner had used copper and excessively large copper ornamented lamps which was not typical of the era. The commission however accepted blame for the error since they had granted approval too quickly, and went no further than to strongly encourage the owner to change the facade again with respect to historic preservation concerns (Historical Commission, File Delaney Street No. 130–132). On the occasion of a later necessary renovation, the house owner voluntarily removed some of the elements which were untypical of the era.

5.5.3.2 255 Pine Street

The corner house at 255 Pine Street was built around 1800. A total of six different tenants and/or owners are documented. In 1957, the Historical Commission recognized it as worthy of preservation and redevelopment, and with the help of the detailed description of the fire insurance company, the RDA restored it as completely as was financially feasible. Subsequently, it was purchased by a physician and his family.

Like the house on Delaney Street, this historical building has only recently been restored to its authentic appearance. In 1990, changes to the facade were made in which, among other things, some existing concrete steps were replaced by white steps in the colonial style, green painted wooden window shutters were added and a new door was installed. In addition, an air conditioning unit was moved from outside to inside the house, the security bars removed and the windows replaced with bulletproof glass (Historical Commission, File Pine Street 255).

5.5.4 Urban living with a Suburban Motif in the Historic District

Society Hill is an almost exclusively residential area, with just as few exclusive shops for last-minute needs. There are also a few cafés on the outskirts. The remaining structures consist of two-thirds historical buildings from the colonial and early post-colonial times and one-third new buildings since 1956. The south and east are directly adjacent to entertainment districts with nightlife and cultural establishments.

The process of gentrification can be observed in many American cities. Society Hill includes houses which are available for rent as well as left vacant. It is notable that the number of vacant buildings increases towards the edge of the neighborhood. The number of demolished buildings (Urban Renewal Areas) and the number for rent and for sale (most of which are still inhabited) decreases towards the perimeter. Society Hill in Philadelphia represents an important example in that it differs in some respects from the other neighborhoods. Within Philadelphia, it is the first neighborhood to have completed the upgrading process.

As part of the revitalization of Philadelphia's CBD, the Society Hill renewal project was received enthusiastically once it was completed. The city planning approach of the 1950s followed the strategy of creating a flourishing economy in the inner city to have a long-term effect on the other districts. It is a fact that many people who otherwise would be living in the suburbs have remained in the city.[124] Still, only a few residents of Society Hill have consciously relocated to the city (Iv-Phl-14). For the gentrifiers, all roads led to the CBD, whereas the poor and socially weak population had to go the way of displacement (Marcuse 1986, p. 155). The socio-economic polarities, as evidenced in the demographic data for various years in comparison, have intensified in recent years.

[124] The thesis of Smith (1979) concerning gentrification.

Society Hill continues to show some of the hallmarks of a gated community.[125] In the context of the Urban Renewal Program, individual street sections were privatized and thereby became an exclusive means of access to the property. Today, these barriers have been lifted, but psychological barriers still exist in the form of chains in front of pedestrian zones hung with signs indicating "Maintained by the residents".

In the context of the discussion concerning the urban-sociological term "gentrification" (Dangschat and Blasius 1990 as well as Friedrichs 1997), the term "anchored tenant" is often mentioned. In contrast to many areas which, like Society Hill, were subject to gentrification, only "unwelcome" population classes have been actively driven away from these areas. New "welcome" residents for Society Hill arrived slowly at first, but options like the building of private streets and other socially selective city planning elements which represent an exclusion of unwelcome visitors have favored the desired influx. In this way, an area is being upgraded within the city while simultaneously creating an outside and inside world. Smith (1979 and 1986a) has also dealt in detail with gentrification in Society Hill. He has come to the conclusion that success would never have been possible without gentrification in Society Hill. Houses restored and renovated in the middle of the city have attracted additional higher-earning residents from the suburbs or young professionals who have just received their college degrees.

Located in the heart of Society Hill is the first black church (Black Church) in the United States. The oldest cemetery belonging to it has been removed as part of the RDA program. Next to this historical church building on the property of the former cemetery, an Afro-American Cultural Museum was supposed to be established in 1973/74. Soon after the announcement of this plan memberships in the citizen organization grew rapidly (Iv-Phl-14). The white upper class of Society Hill had reservations concerning increasing visitors and traffic. The Museum was considered incompatible with the mixed usage of Society Hill (Weiler 1980). A brief but successful campaign prevented the Museum for the ethnic minority from being built in the neighborhood. This successful opposition on the part of the influential neighborhood organization against a museum for the history of African Americans is an example of how the social ecology of Society Hill is promoted. The members do not want any extraneous visitors to their ethnically and socially homogeneous neighborhood (Iv-Phl-1, Iv-Phl-14).

The boundaries of Society Hill, which is to say the creation of a new neighborhood in the context of this project, is a further point of particular significance. The boundaries were drawn in such a way that Society Hill could be presented as a homogeneous section of the city. The office buildings, South Street and the subsidized housing and hospital grounds have not been included in the zoning. The social and ethnic mixture of Society Hill has therefore been predetermined (Weiler 1980, p. 229).

[125] Regarding more recent research, mention should be made here of the lecture by Prof. Frantz (German National Geographers' Meeting in Hamburg, 1999, in the USA Specialty Group).

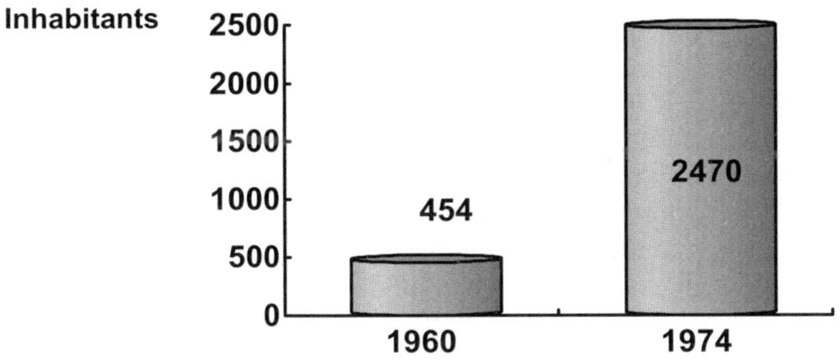

Figure 5.2. Population of Society Hill 1960 and 1974 (Source: Garvin 1996, p. 216)

An increasing problem over the past few years has been parking spaces for the residents of Society Hill (Phil Plan 1980). The individual property parcels are in most cases too small to park a car (Weiler 1980, p. 226). Because the historic dwellings have few stories, no new construction is higher than three floors, thus eliminating the possibility of a high-rise garages. Only dwellings and trees are to be found on the streets. There are a few carparks and parking lots. But every available or soon-to-be available lot is the subject of a debate as to whether it should be used for recreation or parking.

From the beginning, revitalization of Center City, the completion of the riverfront and recreation park Penns Landing and the increase in visitors and tourists, parking has become a highly problematic issue.[126] The neighborhood organization of Society Hill declared itself against making their well-situated (from a traffic standpoint) neighborhood available as a parking lot (Society Hill archive). The city planning office is currently studying various options for sensibly solving the growing parking problem for visitors to the recreational and cultural facilities.

This project was actually supposed to bring back families who had moved to the suburbs. The demographic data, however, show that most households consist of only two persons. The gentrification process of Society Hill has even today not generated any wave of city-wide revitalization. Society Hill's spill-over process has inspired other parts of the city to consider gentrification processes: the Old City district to the north is already showing similar tendencies (Iv-Phl-13; Cybriwsky 1986, pp. 27). Likewise, the massive buying up of entire blocks by the University of Pennsylvania in West Philadelphia fits this pattern. Nevertheless, one can speak here of a success of sorts, since the neighborhood in its present form doubtless contributes to Philadelphia's image enhancement.

As early as 1961, Jacobs (1933) described how city planners paradoxically destroyed city quarters while actually intending to preserve and upgrade them. Gentrification in Society Hill has taken place more from a free market economy point

[126] According to figures provided by the Visitor Center, tourist numbers for 1994 rose by 27.3% to 80,046 (orally conveyed by Mrs. Fisher from the Philadelphia Visitor Bureau).

of view than one based on social considerations (Cybriwsky 1986, p. 53). In their analyses of the ethnic composition of Philadelphia neighborhoods, Goode and Schneider (1994) regret that the social and cultural diversity was not able to be exploited in a positive way. In many respects, Society Hill is similar to any one of the modern suburbs in the Philadelphia region (Cybriwsky 1986, p. 32). Still, it should be emphasized that this project was not explicitly about creating a suburb, but rather intended to restore a vibrant city life in Center City Philadelphia (Iv-Phl-4). The mixed use of apartments, business, office buildings and industrial sites often found in city centers can no longer be found in Society Hill.

To summarize, the extent to which gentrification in Society Hill is unique bears repeating: There is no other example to date in which such a dilapidated city quarter has risen from the ashes into such a well-to-do and intact new quarter.

5.5.5 From National Historic District to Local Historic District

Twelve years later, in March 1999, Society Hill was legally designated by the City of Philadelphia as a Local Historic District. The borders have been drawn somewhat tighter compared with the National Historic District. Among other things, they were kept closer to the neighborhood of Society Hill, and in addition, there are already several new and newer non-historic buildings within the borders of the National Historic District. Work on this project has gone on for several years, carried out by residents and the neighborhood organization of Society Hill, not least because since 1987 the patronage of the RDA ran out in 1956 when Washington Square East (Urban Renewal Project) was started.

The "Plan for Center City" (1988) provides for supporting this plan even if no additional financial means (such as for administrative matters) are forthcoming. Only the Philadelphia Historical Commission, which is responsible for the entire city area, works on the project. In recent years the Commission undertook an inventory which registered every historic house in the proposed district. Even though the neighborhood organization had already presented its own draft, the office in this case insisted on taking its own inventory (Iv-Phl-10, Iv-Phl-14).

Future renovation and redevelopment work is therefore since March of 1999 being funded not only through several state incentives but also using local tax write-off opportunities. The involvement of the neighborhoods is based among other things upon the fact that there is relatively little mobility in the residential population which with its high level of education and above average income, is able to create interest in and raise money for this project. The project will lead certainly not only to a further strengthening of the already existing close neighborhood structure, but also to a new wave of home restoration by the property owners (Iv-Phl-14).

5.6 Historic Preservation as a Perspective on the Future for Philadelphia

One of the most important elements which should help Philadelphia in its historic preservation is strengthening the self- and external image of the city. Philadelphia has been able to clearly expand tourism over the years (PCVB 1997, p. 2). The self- and external image of a city consists of different components which are in part not logically verifiable. Table 5.4 shows some indicators which have markedly stamped Philadelphia's image.
Wailer summarizes the self- and external image of Philadelphia's inhabitants as follows: "Philadelphians lack a strong positive image of their city, their city's government and their own roles as citizens of the city. They have a strong interest in neighborhoods, in their private affairs, and in alternative ways of life to city living." (Weiler 1974, p. 14). This attitude observed by Weiler in the 1970's can also be confirmed in the present time. The following table lists different image improving advertising slogans which were used in connection with various city marketing projects (Table 5.5). These slogans were developed primarily to generate a positive outward and inner image; in some cases direct responses to prejudices can be seen (for example 1970).

Table 5.4. City image presentation through movies filmed in Philadelphia

Year	Name	Image of Philadelphia and reason for selecting Philadelphia
1941	The Philadelphia Story	A love story showing the stubborn, ambitious Philadelphians and Philadelphia during the booming phase.
1971	Rocky	The economical plight of the city mirrors itself in the misery of Rocky (Sylvester Stallone), shows how the working class lives, significance of sports for holding together an urban community.
1985	Witness	The movie is about an Amish boy who is only witness of a murder. Philadelphia was selected here as a showplace for religious tolerance and freedom.
1994	Philadelphia	The City of Brotherly Love as location for a movie that uses AIDS as a call for humanity and tolerance.
1995	Twelve Monkeys	A movie that tells the story of apocalypse in a dilapidated North Philadelphia.
1999	The Sixth Sense	The historic facades of Society Hill serve as a backdrop for the extra-sensory powers of a young boy.

Table 5.5. Improvement of internal and external image: Image improvement and advertising slogans for Philadelphia (Source: Urban Archive (Philadelphia), Files different years)

Year	Slogan	Medium
1947	Better Philadelphia Exhibition	The vision of the future Center City was introduced to the public as part of an exhibition.
1970	Philadelphia isn't as bad as Philadelphians say it is.	Billboard
1976	Yo! Philadelphia	Greeting phrase used for advertising purposes.
1977	The Great Philadelphia Land RUSH! The New Philadelphia Story. Our business is helping your business.	Advertisement in newspapers and magazines.
1978	Join the Posse! I'm looking for volunteers for a posse to catch "Bad- Mouth" and his gang. They've robbed this town of ours too long.	Advertisement in newspapers and magazines.
1980's	Join the Parade of "The Philadelphia Boosters Association" -An Organization Dedicated to Boosting Philadelphia-	Poster
1980's	Clean up Philadelphia!	Inscription on trash cans
1980's	I'd rather be in Philadelphia (By W.C. Fields)	Advertisement
1980's	Philadelphia, Style.. We have got it!	Poster, Advertisement
1981	Move over Pittsburgh and Dallas - Philadelphia's <u>the</u> City of Champions!	Billboard (The pride of the populations city is being directed towards the two professional sports teams)
Ca. 1985	Another Society Hill	Advertisement The successful gentrification process of Society Hill, which was setting an example, should be repeated in other sections of the city.
1988	Benjamin Franklin updated: Join or Die Bucks, Chester, Delaware, Montgomery, Philadelphia	Advertisement proclaiming the necessity of the regional teamwork and dependence.

Year	Slogan	Medium
1990's	Back-to-the-City-Program	The districts should be revitalized as part of financial assistance programs.
Year	Slogan	Medium
1990's	A Clean City is a Neat City!	Inscription on trash cans
1993	Phila-pride	Drawing/painting contest, poster, the project is repeated in 1999.
1994	Make it a Night!	Advertisement for encouraging evening activities in Center City.
1994	Philadelphia - adel - phia	City signs. Refers to the pride of being a city with history.
1995	Phlash downtown!	Advertisement for city tour buses in Philadelphia.
1995/6/7	Philadelphia is on the Move!	Inscription of all the financed renewal projects of the city.
1996/7	Make it a City!	Radio advertisement
	Go and shop in Center City.	Brochures in the shops of Center City.
	Make it a Center City!	
1997/8	Welcome to Philadelphia- Enjoy our Past, Experience our Future	A contradiction which was developed especially for tourism in Philadelphia.
1998/9	Welcome to Philadelphia - The Place that Loves you Back	This slogan was intended to create distance from the purely touristic reference while at the same time aiming for a positive self-positioning of Philadelphians.
1999	Phila-pride	A slogan directed at the population, intended to promote individual identification with the city. Also intended to counter vandalism; based on the first project ideas from 1933.

The image shift which Philadelphia is currently experiencing is clearly a positive development. A real difference will only have been made however when the Philadelphians themselves stand behind the new motto "Philadelphia - The City that Loves You Back". Tourism and the viewing and appreciation of the historic city are still a movement which has not yet reached all the population groups.

The motto: "Philadelphia's Proud Past – and Uncertain Future" which was coined and propagated mainly by the newspaper press will probably continue to apply to the city for some time, and it remains to be seen what value and what economic and social benefit Philadelphia will draw from historic preservation. There are to be sure some "bridge concepts" to other parts of the city, but no social or ethnic mixing worth mentioning has occurred as a result. The many examples of Victorian architecture are not being incorporated into the historic preservation of Philadelphia (Datel 1985, p. 128). In this connection it should be mentioned that the Victorian areas of Philadelphia (especially West Philadelphia, South Philadelphia) are demographically weaker, which is one reason why historic preservation is not as advanced in these areas.

The city planning and historic preservation interaction in Philadelphia is subordinate to the whole. To create simply a "Historiopolis" out of Philadelphia would make no sense, even though the city is already anchored in the minds of the population as a cultural center. Rather, emphasis should be laid from the political side on industry and the expansion of the service sector. This sector guarantees jobs and long-term connections to the region, which in turn would probably have a positive effect on historic preservation activities.

To increase the attractiveness of Philadelphia for investors, a tax reduction must also be introduced. In a tax comparison Philadelphia is even ahead of New York for investors (City of Philadelphia 1999, p. 110). This trend can be seen in all the cities located within the industrial belt. Cities in the sunbelt enjoy comparatively low taxes (see: Appendix B.6). The tourism potential in Philadelphia could likewise be better exploited. Restaurant, room and rental car taxes in Philadelphia are very high compared with other large cities like Chicago, Boston and Washington, DC (City of Philadelphia 1999, p. 111).

5.7 Philadelphia: The Urban Renewal Type

Society Hill, fully planned in its totality, is frequently cited in specialist literature (Fitch 1995, p. 42). As shown earlier, Society Hill should be considered an exception with respect to historic preservation. The new uses of the mixed quarter with warehouses, sheds and old shops and the incorporation of the old historical structure into modern, comfortable dwellings is today an outstanding example of historic preservation in the U.S.A., but we must remember that it was not intended as an historic preservation project. The Society Hill "rehab-project", as Gutschow put it (1978, p. 140), was rejected by historic preservationists at the time. This opinion is now no longer justified, since the critics of those days had not yet realized that the entire neighborhood could otherwise have fallen victim to the wrecking ball. Besides, the initiative of the neighborhood organization has today gained strong credibility (Iv-Phl-14, Iv-Phl-15).

Nearby recreation areas, green spaces (parks), parking spaces and access to shops according to short-term and long-term needs can be classified a success in Society Hill. This success can be attributed not least to the joint venture between public and private institutions, especially the neighborhood organization. The Delaware Valley Regional Planning Commission (in: City of Philadelphia 1999, p. 28) estimates a population increase of 14.2% for Center City within the next few years. A population loss is forecast for all other districts. These numbers make clear the continuing and increasing attractiveness of Center City. An expansion of historic preservation project actions and the further population growth in Center

City will continue to be supported in the future. The "suburban motif" of Society Hill should not be overlooked (Cybriwsky and Western 1982, p. 356).

One should not however overlook the other side of the coin when considering the successes of this historic district: The (re-)creation of preservation-worthy historic colonial Philadelphia has taken place without awareness of the historically authentic buildings from the 19th and 20th century. All buildings which were not constructed during the colonial or immediate post-colonial time were torn down and street fronts erected strictly in the Federal style. Historic colonial style houses were also torn down in the name of so-called improvement to the infrastructure, which meant, for example, that new streets were built. Houses on the river side of the Hill in particular fell victim to this project. Several docks in the harbor had to be completely demolished to build Interstate 95. Up to 16 lanes of highway prevent unrestricted access to the Delaware River to this day. It was for this reason that a pedestrian bridge was dedicated as an extension to South Street in 1998.

The river beds which have been filled in over the years and which could be reopened represent further potential for Society Hill. In addition, the RDA has erected new buildings which are compatible with the other houses in the area only in color, outer facade and height. Only the churches and synagogues which were erected later have been preserved and today characterize the silhouette of Society Hill. Uhlig (1977) describes this process as "style shrinking". Clearly, there are differences between preservation and upkeep, which in this context might be labeled "gray" historic preservation.

Both typical and atypical patterns of city revitalization have been shown here. Philadelphia functions here as an urban renewal type of urban historic preservation in the United States. An important and correct choice was made in the 1950s when it was decided to redevelop the neighborhood in this form. Once the RDA had sold the buildings as part of the revolving fund and the neighborhood demographics had stabilized, the PHC, along with the neighborhood organization, assumed responsibility for quality assurance in the neighborhood. Today, many individuals take over the detailed administrative work, such as verifying whether property owners have observed the historic preservation regulations.

Photograph 5.1. Society Hill with its historic buildings in the colonial and post-colonial style.

Photograph 5.2. The three Society Hill Towers erected in 1964 in the middle of the historic district. They were supposed to represent the first step in the Urban Renewal Program for revitalizing Society Hill by providing modern apartments for higher income classes.

Photograph 5.3. The new buildings fit in well visually with the historic image of Society Hill (Delancey Street between 4th and 5th Street).

Photograph 5.4. A "touristy" atmosphere in the style of "living history" in Society Hill (Pine Street between 2nd und 3rd Street).

6 Historic Preservation in Boston

6.1 City history of Boston

Boston is the capital of Massachusetts and is located on Massachusetts Bay. The name Boston derives from St. Botolphs Town, which goes back to the founding of the city by the Puritans.[127] With 574,238 inhabitants, Boston is a large city (U.S. Census 1990). The area of the City of Boston consists of only 48.8 square miles, which, compared with other cities in this empiric investigation, gives a high population density (Appendix B.1).

In 1630, Boston was initially founded, where Salem is nowadays located, but soon after, the city was established on the peninsula of Massachusetts Bay. The religious and economic life of the Bostonians was marked mainly by the presence of the Puritans. Today's mannerisms and economic structures can be traced back to this religious attitude. Baltzell (1980) has made it the subject of his book "Puritan Boston and Quaker Philadelphia" to depict the differences of the two cities with regard to the city's foundation, constitution, work ethics and productivity as well as innovative thoughts.

In earlier times, Boston was mainly a commercial center for wool, leather and fish. Many products were imported to the colony from England, prior to the Declaration of Independence in 1776 (Sautter 1986, pp. 146). Due to the early foundation of academic institutions, Boston has also become a publishing center. In the 18th century, the New England province developed in such a way that a take over by the Bostonian residential businesspeople was boosted as a result of economic and political factors. These businesspeople took over the leadership of the struggle against British policies, which reached a peak with the Boston Tea Party: In 1773, a complete shipment of tea from the English "East Indian Tea Company" was thrown overboard by businessmen who had disguised themselves as Mohawk Indians (members of a secret organization, "Sons of Liberty") (Dahms 1983, pp. 46 and also Sautter 1986, p. 541).[128]

---

[127] The common nickname "Beantown" derives from the colonial days as well. At that time beans were a significant commodity and foodstuff.

[128] An interesting aspect of this piece of American history is that the object in question is not a building, but rather a ship, which was supposed to be placed under protection. The original ship no longer exists. A replica in the harbor is the scene for regular reenactments of the Boston Tea Party.

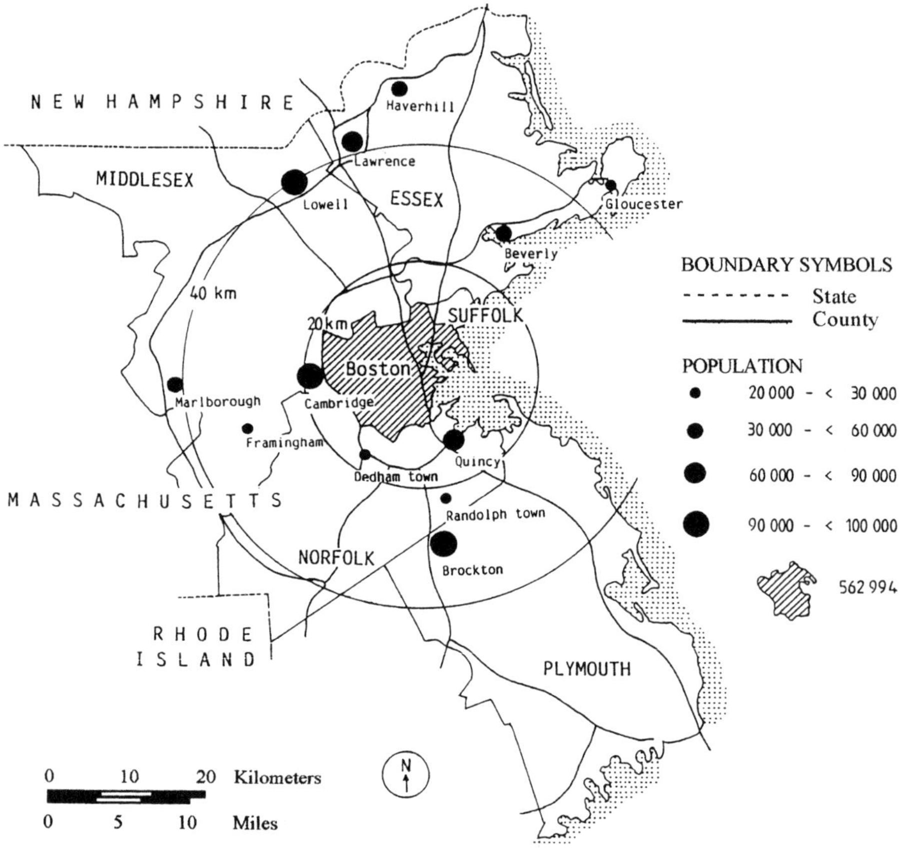

Map 6.1. Boston SMSA (Source: Hahn 1990, p. 27, modified)

After independence, Boston was made the center of the China trade, shipbuilding and American shipping. With the first publication of the magazine "The Liberator" in 1831, Boston became the center of opposition to slavery.

Today, Boston has several nationally and internationally recognized universities, archives, museums and theatres as well as the headquarters of large companies. In addition, the harbor is one of the major U.S. Atlantic ports, which also explains the importance of shipbuilding for the city and the region. The significance of Boston extends far beyond the consolidated SMA of about 5.4 million inhabitants.[129] It is also of interest that 600,000 people are employed in Boston, a figure exceeding the actual population (BRA 1995b, p. 2). The gross national product of the city is derived mainly from (in descending order) banks, health care, education/training, technology and tourism (based on U.S. Census 1990).

[129] This makes Boston the seventh largest Statistical Metropolitan Area within the USA (BRA 1995b, p. 2).

6.2 Historic Planning Districts and Building Structures

A complete mapping out of Boston's architecture has been undertaken by the AIA (Southworth and Southworth 1992). The authors divide the cities of Boston and Cambridge into 16 different architectural units. These are derived from the Boston Redevelopment Authority (BRA) and serve as the basis for organizing planning and research projects (BRA 1995b, p. 5). Boston's traditional neighborhoods were taken into consideration when drawing this map. Along with these divisions there are an additional 64 so-called Neighborhood Statistical Areas, which represent the sub-areas of the 16 BRA-designated neighborhoods. One of these 64 NSA is Beacon Hill, which will be examined in greater detail below.

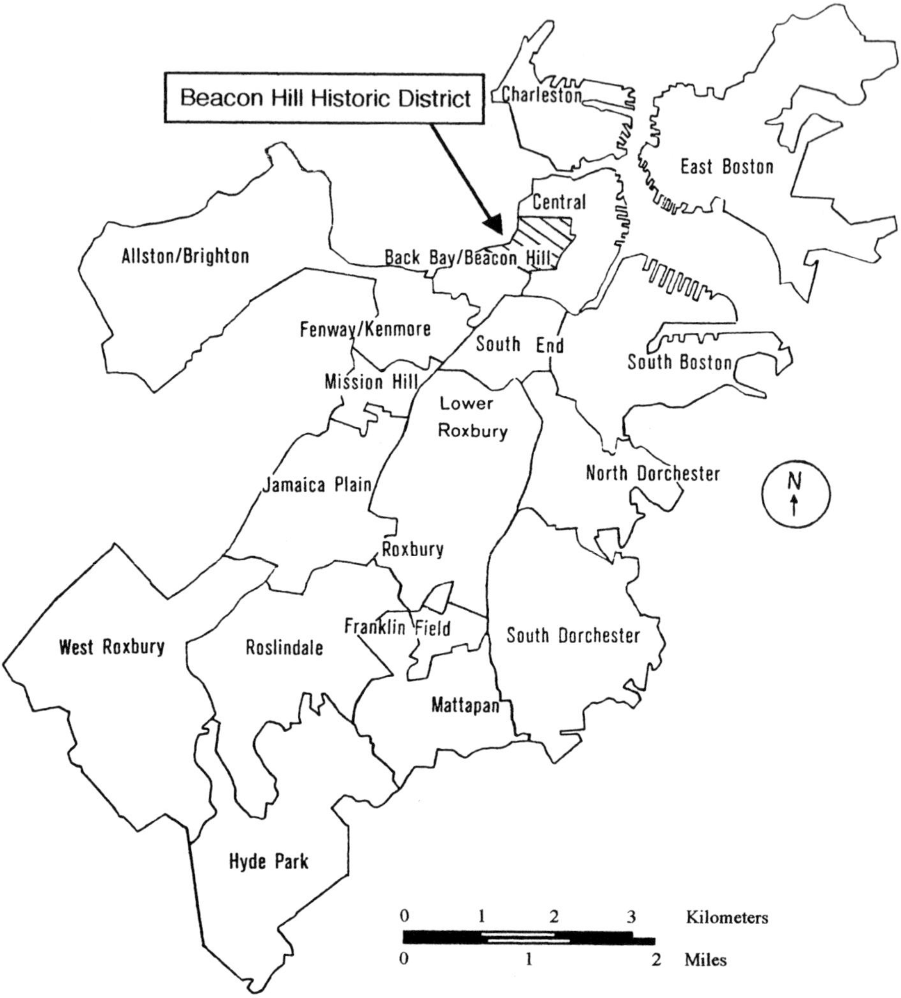

Map 6.2. City of Boston, location of Beacon Hill (Source: Schneider-Sliwa 1995, p. 587, modified)

The Beacon Hill Historic District (Map 6.3) consists mainly of buildings from the time between 1800 and 1850. Federal Style, Gregorian Style, Greek Revival and Victorian buildings predominate in Beacon Hill, with a few townhouses having an architectural mix not associated with a particular era. Table 6.1 documents the demographic data of the Beacon Hill Historic District compared with the City of Boston.

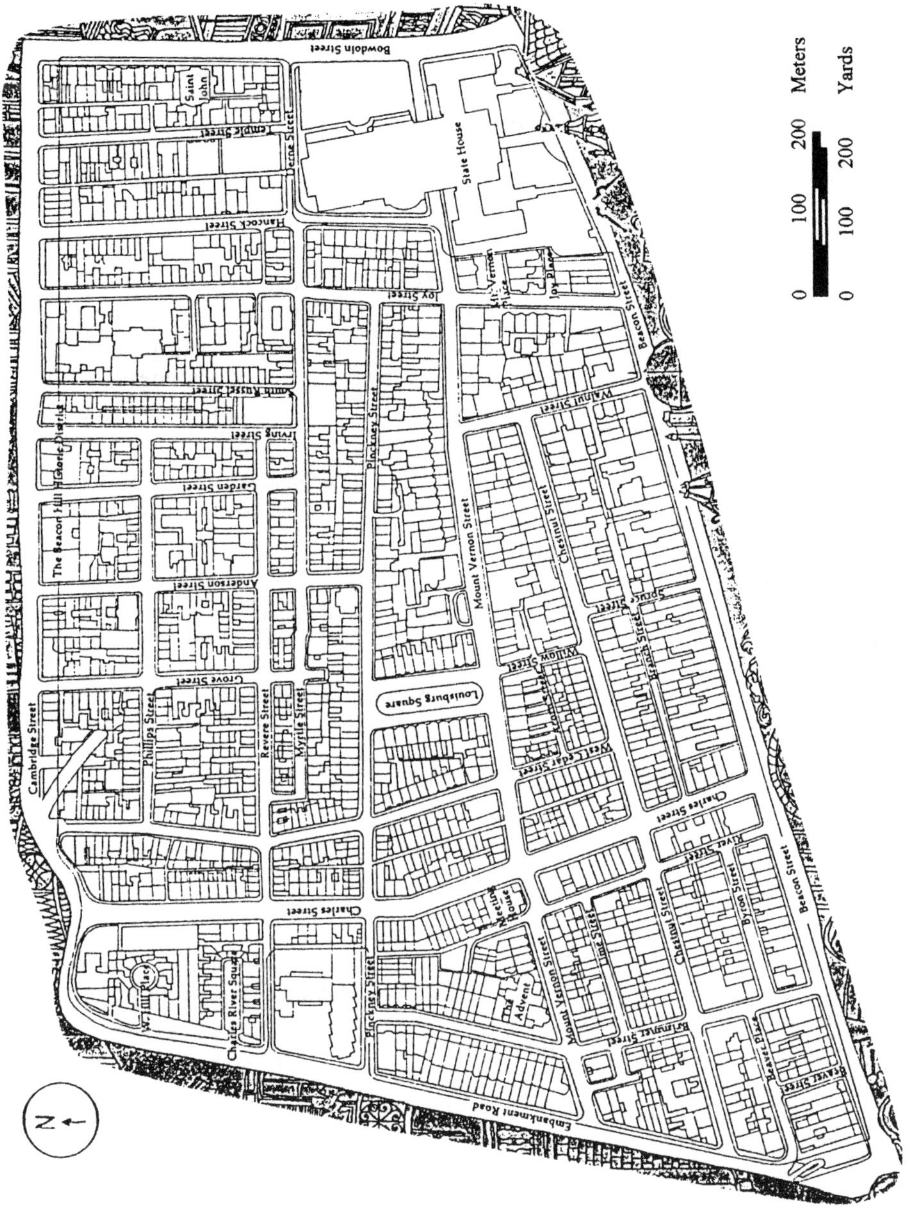

Map 6.3. Beacon Hill Historic District (Source: Beacon Hill Civic Association (BHCA) 1975, Appendix)

Table 6.1. Structural data of the City of Boston and Beacon Hill Historic District compared

	City of Boston	Middle Value (in percent)	Beacon Hill Historic District	Middle Value (in percent)	Comment/Evaluation
Total population	574,280	100%	9614	1.7%	--
White	361,513	63.0%	9039	94%	Very high percentage of white population in Beacon Hill
African American	146,695	25.5%	194	2%	Very small percentage of black population in Beacon Hill
Hispanic Origin	59,692	10.4%	126	1%	This group includes Puerto Ricans etc.
Minority Status: All minorities[a]	234,825	40.9'%	651	6%	In comparison to City of Boston, the minority status in Beacon Hill (only 6%) is rather low.
Average Age	30.4	Largest value lies between 25 and 34 years (23.2%)	32.9	Largest value lies between 25 and 34 years	In Beacon Hill, there are few very young and very old inhabitants; the average age is equally high in all groups however.
Persons over 65 (all ethnicities)	66,156	11.5%	771	8%	There are fewer old people in the historic district
Number 0-4-years old (all ethnicities)	35,772	6.2%	228	2.4%	There are fewer babies/small children in the historic district
People living in families[b] (all ethnicities)	388,092	67.6%	3655	38%	In Beacon Hill, there are fewer families
Non-family households	153,185	26.7%	5708	59.4%	There are essentially more people living in the historic district who are not family-bound

	City of Boston	Middle Value (in percent)	Beacon Hill Historic District	Middle Value (in percent)	Comment/Evaluation
Number of single-parent households (all ethnicities)	46,566	20.4%	177	1.8%	There are hardly any single parents in the historic district
Number of single-parent households (African American)	20,097	59.3%	10	0.01%	Disproportionately small numbers of African American single parents live in the historic district
Work-status: Non-working women[c]	86,568	18.2%	787	0.82%	There is a smaller percentage of women working in Beacon Hill compared with the city of Boston overall
Unemployment rate	--	8.3%	--	5.2%	Unemployment is lower compared with the City of Boston
Median income by household (all households)	--	29,179 US $	--	38,089 US $	Clearly higher income in the historic district Beacon Hill
Person receiving public assistance (all ethnicities)	80,541	14.9%	103	1.2%	There are essentially fewer people receiving state social support, most of them are families with children under 6 years
Education-level: Number of people with College-degree	109,710	30.0	5634	58.6%	The level of education in Beacon Hill is clearly higher

	City of Boston	Middle Value (in percent)	Beacon Hill Historic District	Middle Value (in percent)	Comment/Evaluation
People living in poverty[d] (all minorities)	--	18.7%	--	10.7%	Conspicuously high percentage in Beacon Hill. The following statistical distortion may be present: family income or support were not included in this statistical evaluation
Rented housing units (total)	250,857	--	6613	--	Conspicuously high number of tenants in Beacon Hill
Average age: Owner	51	--	47	--	The average age of house owners is lower than in comparison with the city of Boston
Average age: Tenant	38	--	32	--	Tenants are younger on average
Living units of each building	--	Average: 3	--	Average: 4	More living units in each building (the buildings were "condominimized"), a similar trend is seen in Charleston
Average age: Buldings[e]	--	1939	--	1939	It is striking that the average age of the buildings is the same as in the City of Boston
Number of buildings built before 1940	144,604	--	5797	--	The number of buildings built before 1940 is similarly high in all areas[f]
Total number of buildings	250,869	--	6613	--	--
Average year of the last move: Owner	1978	--	1984	--	--
Average year of the last move: Tenure	1987	--	1989	--	The renting period in Beacon Hill is very short, a sign of gentrification

	City of Boston	Middle Value (in percent)	Beacon Hill Historic District	Middle Value (in percent)	Comment/Evaluation
Average Rent	--	625 US $	--	765 US $	--
Tenants according to ethnic origin: African Americans	38,300	--	78	--	Only very few African Americans rent apartments in Beacon Hill
Number of Housing Units by house: 1 Bedroom (BR)	53,868	--	1779	--	1BR-Apartments predominate in the historic district
1BR - Average rent	570 US $	--	760 US $	--	Cleary higher rents in comparison to the City of Boston

[a] This statistical group includes all above mentioned ethnic groups including Native Americans, Asian-Pacific Islanders and "others".

[b] The statistics refer to traditional family units. Unmarried couples with children are not included in this category and affect the other categories.

[c] All employable persons over 16 years of age.

[d] In 1993, the poverty line was defined as follows: The upper income level for a family of 4 is 14,763 US $ (U.S. Census 1995).

[e] A separate assessment for the age of buildings in Beacon Hill and Back Bay can be found in a later table.

[f] The U.S. Census makes no further differentiation concerning the age of buildings before 1940. This historical data can only be found in individual assessments through the Boston Landmarks Commission (BLC). Special assessment according to the U.S. Census 1990.

6.3 Beacon Hill: Local Historic District

6.3.1 Beacon Hill: History and Roots of Historic Preservation

Beacon Hill is a neighborhood within the core city area of Boston. In 1955, Beacon Hill was declared a National Historic Landmark. The historic preservation being practiced there is often put forward as an exemplary model in the USA.
Historically and architecturally, Beacon Hill has always been viewed as a distinct residential neighborhood whose viability today results in large part from a history of sound planning, adaptability to change and well-organized community groups. In the 18th century, Beacon Hill was for two decades the seat of the state government of Massachusetts, located in the Charles Bulfinch[130] house (Southworth and Southworth 1992, p. 144).

The flattened hill upon which the neighborhood is located is where the first European settlers arrived, making this part of what later became Boston their home. The micro-topography was actually different at the time, consisting of three uneven individual hills originally known as Trimountain. One of them, Mount Vernon, was 1.80 m higher than today's highest point, Louisburg Square in Beacon Hill. The three peaks were soon cut down to make building on the South Slope possible and start the filling of the Boston shore and river lines. Many buildings were then rebuilt or restored, since they been built mainly of wood. Due to a lack of space and construction types there were no front yards in Beacon Hill. Many houses have a tunnel-like walkway to the backyard, a building code of 1830 having specified that access measuring the width of a cow and height of a young boy with a basket on his head be provided.

The streets in Beacon Hill are laid out in a grid. However, the individual streets vary in width and length and several have slight curves. With only three through-streets and a complex one-way system, traffic is discouraged from using Beacon Hill as a short cut, which would otherwise mean considerable commuting traffic.

The attraction of Beacon Hill is due to several factors, chief among them the idyllic and well cared for atmosphere of the district (BHCA 1975, p. 1). Entering the neighborhood, one gets the feeling of leaving modern-day America. An historic, untouched world preserved down to the smallest detail greets the visitor (Southworth and Southworth 1992, pp. 144). Many people associate the inhabitants of Beacon Hill with rich Boston Bohemians, even though in fact the population of Beacon Hill has always been very mixed. Especially noticeable is the diverse planning and architectural character of the North and South slope of Beacon Hill. The majority of North Slope residents are young workers and students. Beacon Hill was a popular place to live from 1870–1880, until the Back Bay, lying further to the east, was planned and developed. The declining popularity of the neighborhood lasted only a few decades. In 1922, the Beacon Hill Civic Associa-

[130] Charles Bulfinch was a famous architect of the 18th/19th century.

tion (BHCA) was founded, a neighborhood association which rallied strong community feeling for the Hill. Even at that time, it was clear to the residents that Beacon Hill was in danger of losing its historical and architectural coherence.

After a short phase, in which the Back Bay got the better of the Beacon Hill neighborhood, the south slope lost a number of families to the Back Bay. At that time, and especially in 1950, one could speak of a Beacon Hill renaissance. Even today, both neighborhoods are close competitors in terms of tourist numbers, public support and class-specific buying behavior.

6.3.2 Beacon Hill Historic District: the Neighborhood Initiative

In 1955, the majority of the area of Beacon Hill was declared an historic district by the Massachusetts General Court. Its establishment in 1955 was due mainly to the initiative of the Beacon Hill Civic Association (BHCA). The Boston district was so successful that it was expanded twice in later years. The Beacon Hill Architectural Commission (BHAC) exercises strict architectural controls. The BHAC is an independent committee which works together with the Boston Building Department.[131] Every modification has to be presented to the Commission. The law which the BHAC applies is the Beacon Hill Historic District Law (Chapter 616 of the Acts, as amended), with the following objective: ".. educational, cultural, economic, and general welfare of the public through the preservation of the historic Beacon Hill district, and to maintain said district as a landmark in the history of architecture." (BHCA 1975, p. 1). This law strongly structures and mechanizes control through the neighborhood organization. The City Law Department is, however, judicially in charge of ensuring compliance with the regulations.[132] In 1958, the historic district was extended to include Lower Beacon Hill, between Charles Street and the Charles River Basin, and the law was strengthened to include the control of the demolition of dwelling units and exterior color (BHCA 1975, p. 4).

In 1963, the historic district was extended again, ".. as far as practicable and to the best of our ability .." (summary from a letter from the BHCA and BHAC dated 3/2/1963 from: BHCA archive) to protect the historical integrity of the neighborhood.

In the same year, the Beacon Hill historic district was designated by the National Park Service as one of the first registered National Historic Landmarks (U. S. Department of the Interior 1996). The neighborhood organization itself uses the following characterization: "For the benefit of all, Beacon Hill is very restrictive in many ways" (BHCA 1975, p. 5). In this context, they note various consulting committees which are available to property owners free of charge. The "Beacon Hill Architectural Handbook" (BHCA 1975) contains all the current guidelines, addresses and application procedures which are of significance to the owners. The

[131] As opposed to other cities, Boston has its own Architectural Review Boards (ARB) for different districts.

[132] Fines of up to $50 per day can be levied for violations.

BHCA is especially sensitive to changes in property use, with particular reference to the parking problem and changes to buildings. New liquor licenses are generally not approved or are even ".. vigorously opposed" (BHCA 1975, p. 8) out of fear that the clientele will disturb the neighborhood.

6.3.4 Maintaining: Urban, Historic, and Residential Harmony

The initial idyllic impression given by Beacon Hill is deceptive, as a view behind the scenes shows: the upkeep of the homogeneous street appearance of Beacon Hill is only possible through the constant involvement of neighborhood organizations. The brochure for homeowners describes how important the unity of the neighborhood community is: "The feeling of neighborhood is more important than the appearance of any single building – including commercial buildings" (BHCA 1975, p. 1). Indeed, Beacon Hill has an unusually active neighborhood organization. Within Boston, the Beacon Hill Civic Association is often dubbed the "Hysterical Commission" (Iv-Bos-5).

In the 1960s, even before design guidelines existed, non-historic authentic streetlamps were replaced by gas lamps. In 1982, a neighborhood initiative was successful in cutting off Charles Street so that it no longer served as a traffic artery and access road (Bostonian Society/File Beacon Hill). Many successful campaigns against heavy city and tour bus traffic have been waged (BHCA archive). Fast food chains have also been prevented from setting up shops in the neighborhood. Such establishments generate a great deal of trash which was seen to pollute the neighborhood. In the context of social welfare programs, the neighborhood called for lower rents, especially for families and older people (Bostonian Society/File Beacon Hill).

The neighborhood regularly files objections against its own neighbors before the BHAC for unapproved modifications to structures. How diametrical the arguments concerning the historic preservation can be, may be seen in the daily press.[133] On the one hand, the neighborhood organization is praised because it safeguards the historic, irretrievable past (Iv-Bos-1), and on the other hand, the overzealous activities of the neighborhood organization is denounced (Bostonian Society/File Beacon Hill). Often it deals with trivial matters, for example the color of the door or the height of the fence around the yard (Bostonian Society/File Beacon Hill). Efforts to protect the neighborhood quality of life often become quite emotional (Iv-Bos-3, Iv-Bos-5). One article shows the political, social and planning power of the BHCA with the title "You can't paint your door on Beacon Hill" (Bostonian Society/File Beacon Hill, Author: Tom Verde).

Nevertheless Beacon Hill has for some years opened up to tourism. In 1998, the Beacon Hill Heritage Trail was created through an initiative of the National Park Service. An illustrated folder guides visitors through the significant historical

[133] A source is here the biweekly: "The Beacon Hill Times". See the January 11th, 2000 Edition (BHAC Report, p. 19).

structures and locations in Beacon Hill of interest to African Americans. Groups or guided tours generally require a permit, however (Map 6.4).

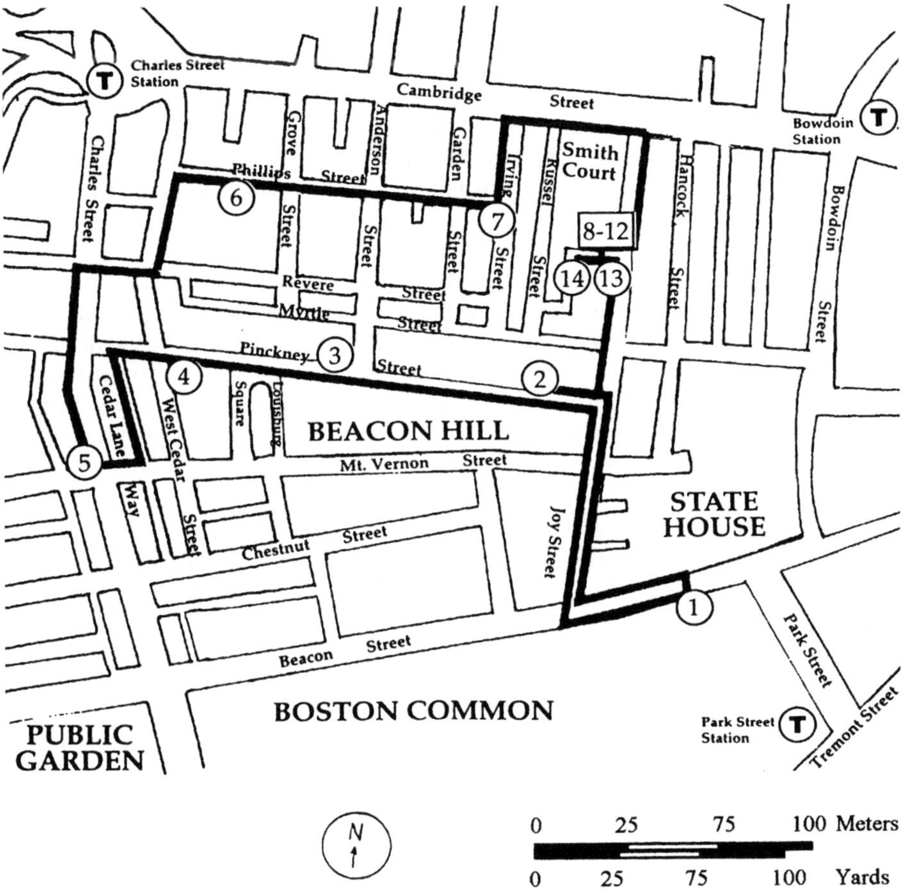

LEGEND

1 Robert Gould Shaw Memorial	5 Charles Street Meeting House
2 George Middletown House	6 Lewis Hayden House
3 The Phillips School	7 Coburn's Gaming House
4 John J. Smith House	8-12 Smith Court Residence

Map 6.4. The Black Heritage Trail within the Beacon Hill Historic District (Source: National Park Service (NPS) 1998, modified)

6.4 Boston: The Neighborhood Type

Since 1955, Beacon Hill and its historically significant structures have been protected through its designation as an historic district. It is clearly evident how actively the neighborhood organization maintains the district and this is an unusual phenomenon. Rutz (in: ILS 1999, p. 30) shows that participative planning in the United States is very effective and in no way marginal. The origins of the historic district in Beacon Hill can be seen as a particularly successful example of this. Yet even after the historic designation and in spite of the historic structures, it today remains a great challenge to maintain the vitality of the neighborhood and make it accessible to tourists.

The competition in Boston between Back Bay and Beacon Hill and, more recently, also with South End cannot be overlooked. Comparisons are frequently made and the historic architectural structures improved. Several citywide, hotly contested historic preservation competitions are held. There is also competition for public funding. Beacon Hill and Back Bay are two neighborhoods which have never been seen as "unmodern" or economically disadvantaged for people moving into the area. Based on previous experience, this makes the initial conditions for the establishment of a National Historic District considerably simpler. This is confirmed not only by analytical observation of the neighborhood, but also by Beacon Hill's demographic data (Table 6.1). Similar gentrified demographic data were shown in the previous chapter for the Society Hill area of Philadelphia (Maps 5.4 to 5.6).

The entire phase of revitalization through Urban Renewal Projects required in Society Hill was absent in Beacon Hill. Neighborhood's structures did not have to be separated completely from the city structure or newly constructed. In Society Hill, the neighborhood structure had been completely destroyed by a combination of the suburbanization phase and the Depression era, so that external renovation was necessary. These existing neighborhood structures, together with the massive support provided by the local city authorities, have contributed greatly to the picture of Beacon Hill which we see today.

Photograph 6.1. The Beacon Hill Historic District (View north from Charles Street)

Photograph 6.2. A private street in Beacon Hill. The exclusionary intent is indicated by a sign "Private Way"

Photograph 6.3. Charles Street in the Beacon Hill Historic District is a busy shopping street with shops of every kind. Even logos for store and restaurant chains have to be adapted to the regulations of the historic zoning ordinance

Photograph 6.4. Since the 1990s, the neighborhood has opened itself up to tourism. Here, as an example the Black Heritage Trail (Joy Street)

7 Historic Preservation in the South: Charleston and Savannah

The scientific reasons as to why the historic districts in the South coast area of the U.S.A. should be taken into particular account within this study are as follows: in these areas, we can identify other examples of architectural structures; but even more significantly (as opposed to the North), the motives behind historic preservation differ greatly. These will be pinpointed in the following analysis. The reason why the North and South have developed so differently is a consequence of their individual history.

The Civil War[134], which took place from 1861 to 1865, was triggered by differences between the northern and southern states (Sautter 1986, pp. 217). These were mainly the result of economic and political considerations and culminated in the issue of slavery (Dahms 1983, p. 250). Eleven southern states left the Union after Lincoln's election in 1860. The northern states insisted that the Union could not be dissolved (Dahms 1983, p. 131). Charleston and Savannah belonged to two Confederate states in the South. This conflict ended after the defeat of the southern states, due mainly to the northern states' material advantages (Dahms 1983, pp. 127-131). The defeat at Gettysburg on 3rd July, 1863 thwarted attempts to divert the war to the North. The American Civil War had taken its toll on society and the economy, and was a great setback to the South. It was not until ten years had passed that the South, following the Reconstruction period (1865-1876), was once again able to liberate itself from the North's political, economic and social dominance (Dahms 1983, pp. 132 and p. 251).

Both Charleston and Savannah contain historic districts within historic inner-city areas. In recent times, the population has been growing steadily within both cities' SMSAs. Within the cities themselves, population growth remains stable (particularly in the historic districts), and, in some cases, the population has actually decreased (extract from the U.S. Census of 1995, see also Fitch 1995, p. 41). Until the 1960s, few historic districts existed in the U.S.A., though they were being introduced increasingly frequently in the wake of a growing awareness of the environment and the image of the community. Here, the southern U.S. cities acted well before their northern counterparts (Charleston, New Orleans and Richmond), as the North only introduced such schemes on a wide scale during the 1970s (Hodder 1999, p. 439).

---

[134] In specialist German literature, this is also referred to as the "War of Secession" = American Civil War (Sautter 1985 and Dahms 1983).

Silver (1991) identifies and summarizes the various stages of historic preservation in the South as follows:

Stage 1. (1920–1950): The foundations were laid for measures to preserve the neighborhoods, demanding a strategy that united the diametrically opposed ideas of planners and preservationists.

Stage 2. (1950–1970): Preservation of contested neighborhoods (demolition or preservation),

Stage 3. (1970–1990): Preservation and historic preservation seen as the key to revitalization.

Today, we can identify a fourth and new stage in which historic preservation is a vital means of ensuring that a particular area or place maintains its specific symbolic worth and enables processes of identification (heritage motive).[135] Projects aimed at developing tourism and the Affordable Housing Programs play a particularly significant role here.

7.1 Charleston

7.1.1 Charleston's Urban History

The first European settlers who moved to take up permanent residence in the area came over the Atlantic Ocean from England on the "Carolina", landing at a nearby bay. There, in 1670, they founded what was to become Charleston as we know it today (Ripley 1996, pp. 9). The harbor was known from then on as Charles Towne Harbor, and the name evolved over time into the modern-day name of "Charleston". Charleston city is built at the precise spot where the Ashley and Cooper Rivers converge.

[135] See Treinen in: Atteslander and Hamm (1974).

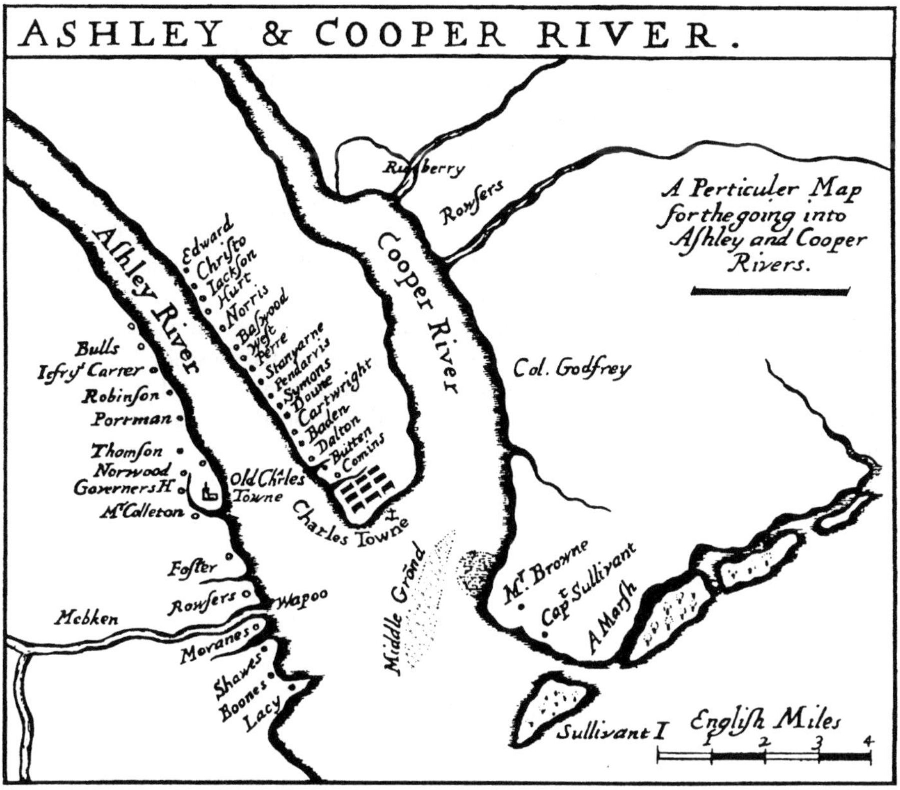

Map 7.1. Charleston city circa 1680 (Source: Ripley 1996, p. 23)

It was vitally important for the first settlers to find a location that could easily be defended, since in the initial years of settlement, outside attackers, namely the Spanish, posed a serious threat. The latter had, at that time, already taken over parts of Florida. The modern-day downtown area (CBD) occupies an area of some 20 km^2. Most residents of Charleston do not live on the peninsula, but in the West Ashley and James Island districts (extract from the U.S Census of 1995). Both districts contain neighborhood areas in which we may find both old and new buildings and urban structures (Ripley 1996, pp. 32).

Originally, only the central part of the peninsula on which the old city was founded lay above sea level (Ripley 1996, p. 49). As the population continued to grow, the outlying areas leading down to the water were subsequently cleared in order to make them habitable. In the period between the early years of settlement and 1850, Charleston had become a wealthy city. Maps clearly show us that the historic nucleus was inhabited early on, and that the population settled in the west and moved eastwards (Ripley 1996, p. 47). Several fires (1740, 1778, 1796, 1861 and 1838) burned many wooden structures to the ground, leaving only the foundations (Ripley 1996, p. 42). In many cases, they were subsequently demolished. Moreover, several natural disasters, such as hurricanes (in 1893 and 1911) and an earthquake in 1886, decimated the cities' structures, damaging them irreparably. In 1860, the American Civil War broke out in Charleston (Dahms 1983, p. 125). Fort Sumter was shut down and federal troops eventually capitulated. The economic and social consequences of the American Civil War were dire and Charleston only truly began to deal with them during the Reconstruction period (Sautter 1986, p. 230).

Since the beginning of the 20th century, we have seen a general move away from agriculture towards trade and industry. Many buildings were lost forever and the population soon became aware of this loss, further reinforcing the historic preservation cause (Dahms 1983, pp. 138).

The street system within the city center area shows all the hallmarks of the U.S. grid system. Due to an acute lack of capital that persisted into the 20th century, Charleston's historical structures remained more or less intact. At that time it considered more profitable, if "undesirable" or "old-fashioned", to use or convert old buildings rather than construct new urban structures.[136]

[136] In the 1960s and 1970s, Stade (Lower Saxony) experienced a similar phenomenon.

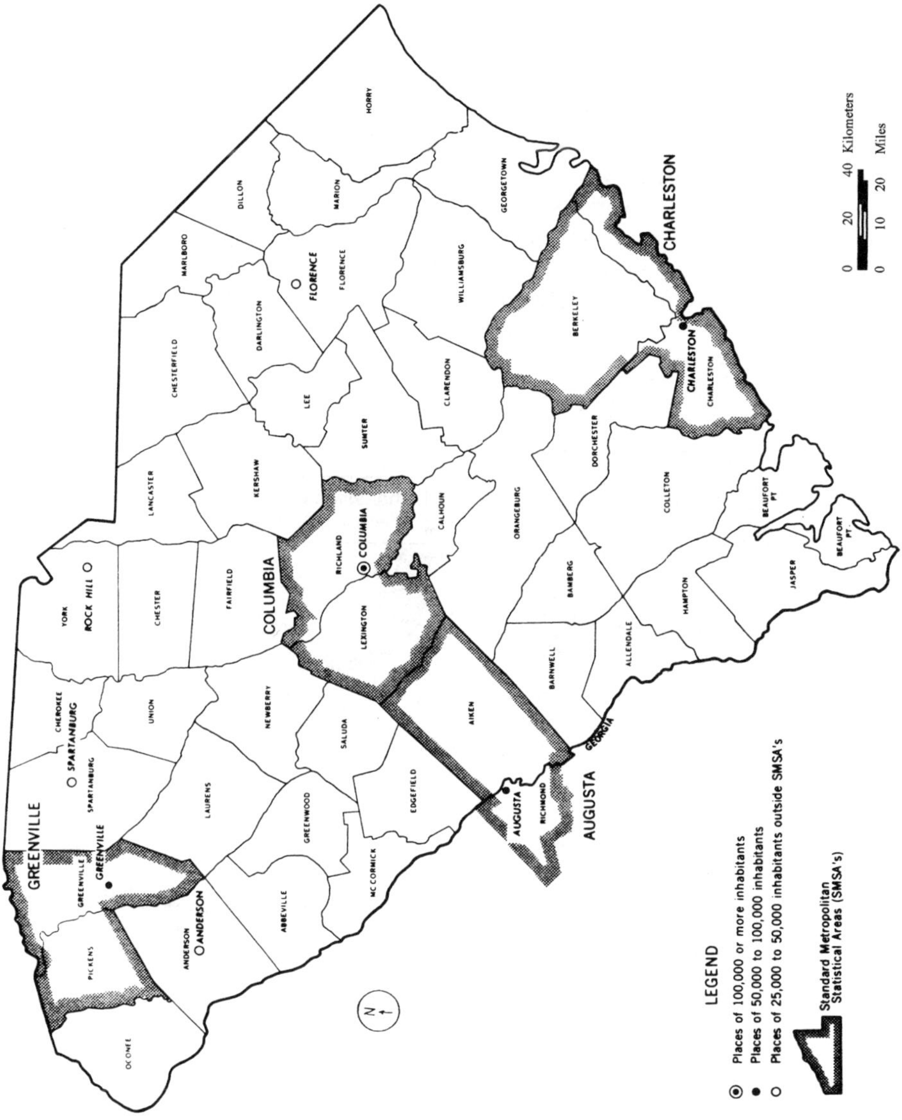

Map 7.2. South Carolina and Charleston's SMSA (Source: Andriot 1979, p. 562)

Between 1980 and 1990, Charleston's population increased by about 15%, whereas it increased by only 9.8% in the U.S.A. as a whole (U.S. Census 1980 and 1990). The city's growth led to further areas in the western and northern parts of the city being reclaimed. In 1998, the port of Charleston had 100,122 inhabitants. Moreover, there is an increasingly common tendency in the U.S. to move to the Sunbelt.[137] The city is, however, attempting to ensure that they do not merely attract retired people and workers who are not active in the area. Financial support programs, development of the study options available at Charleston College and changes to the zoning ordinances to allow buildings to be used for commercial purposes have been introduced to attract other social groups to the Charleston region.[138]

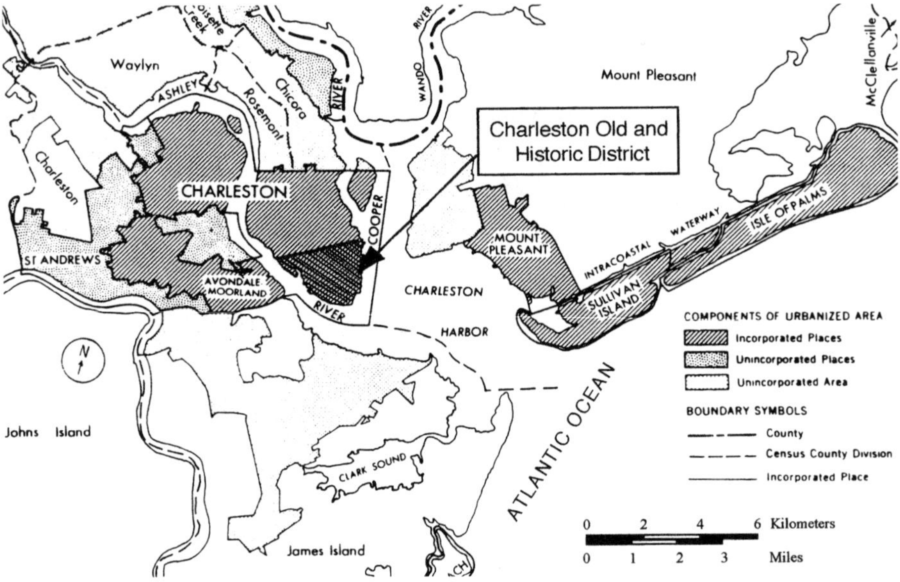

Map 7.3. City of Charleston (Source: Andriot 1979, p. 570, modified)

[137] This is due, above all, to the climate, though Charleston's recreational facilities also attract people to the region.

[138] In 1999, Atlanta, GA, was, the fastest-growing region in the United States.

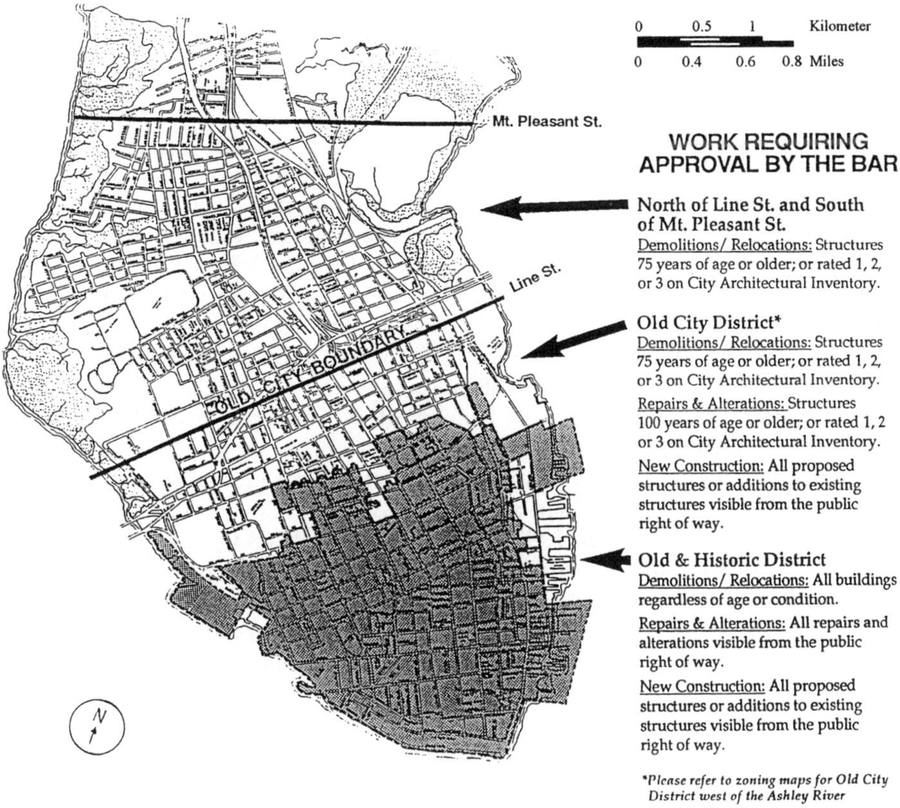

Map 7.4. Charleston Old and Historic District (Source: City of Charleston and Board of Architectural Review 1996, modified)

The city's image was also significantly affected by the foundation of a military academy in 1842. Charleston harbor has always been important for coastal and overseas trade. Until about 1850, the rice, cotton and indigo trades thrived. Naval construction and, later, also engine construction works were a vital part of the economy (Ripley 1996, pp. 30). Today, the food, textile and chemical industries, the crude oil refinery and the tourism industry are at the economic heart of the region. In 1980, some 2.1 million tourists visited Charleston, and 15 years later (in 1995), this number had risen to 5 million. The 1.5 billion U.S. dollars spent by these visitors to the region is an important economic factor (extracts from the U.S. Census of 1980 and 1995).

At the present time, attempts are being made to make Charleston's SMSA more attractive to tourists, motivated by recessive economic tendencies.[139] The program focuses on beautifying buildings occupied by commercial enterprises and companies in the service sector, as well as the secondary and main branch offices of

[139] Charleston Consortium meeting, 20th May, 1999.

large companies.[140] On the other hand, Charleston's Internet site claims it to be a "livable and walkable city", a concept that is largely unheard-of in Europe and Germany.[141] In the U.S.A., it is commonly said that "cities are not inhabitable", the suburbs being seen as preferable. The city is aiming to attract a new category of qualified and financially secure inhabitants. Such attempts are relatively successful in the United States, as Americans are more willing to move around and are not so personally attached to an area as their European counterparts. A large proportion of the U.S. population (can) even choose their home and workplace to suit their leisure activities. Here, however, some features inherent in Charleston's image may deter investors. The aura of peace and quiet and the advantages of Charleston's historic districts (and not merely for tourists) are difficult to combine with the programs implemented to clear land for the profitable construction of new buildings and Charleston's high privatization rate.

The historic district's desire to be part of a "walkable city" (quoted from Charleston city tourist information) means that shopping areas, urban furnishings (such as traffic lights, sidewalks and park benches) and the traffic network have been constructed with pedestrians in mind (AIA 1965, Bloomer and Moore 1975, Lynch 1982 and 1997, Cohen 1999). This image goes hand in hand with aesthetic concerns to preserve the immediate area, of particular significance within the historic districts. The crime rate is relatively low due to police presence and citizen initiatives. The virtues of Charleston's "walkable city" are effectively marketed as an "extraordinary experience" to its inhabitants (Iv-Cha-1).

7.1.2 The Roots of Official Historic Preservation

The Society for the Preservation of Old Dwellings was founded in 1920 by the real estate agent, Susan Pringle Frost, and some of her associates who were concerned to assure the future of the Joseph Marigault House. Frost had already bought and, as a result, saved other historical houses (Rhoad 1994, p. 13 and Silver 1991, p. 70). The house was completely renovated through private donations and, in 1928, became Charleston's first museum. The society developed into a movement that set up similar societies throughout the United States (Bland 1994, pp. 141).[142] In 1950, the society was renamed the Preservation Society of Charleston in order to include wider aspects of historic preservation. Bland (1994,

---

[140] In this case, the Charleston Regional Development Alliance (an organization that brings together public and private representatives) is attempting to develop new concepts. Of particular interest is that major factors, such as corporate identity and historic preservation, play no significant role here.

[141] http://www.charleston.com (15th May, 2000).

[142] Susan Pringle Frost's life was discussed in detail by Bland (1994) in his monograph. It is impressive just how much the historic preservation movement in Charleston owes to this one woman. She is portrayed as both a modern woman of the South and a person so far ahead of her time that she seems almost to have travelled back in time from the future.

p. 180) comments that at a time when women had so few means at their disposal, Frost nevertheless achieved much within the field of historic preservation, a field whose significance was not yet widely accepted. In this context, we should mention that women were also instrumental in carrying out programs of historic preservation in both Boston and Philadelphia.

During this same period, the Charleston city council formed a Special Committee on Zoning in 1920. For the first time, permission was refused to carry out conversion work on South Broad Street, where many historical buildings were to be found. Early on, therefore, Charleston authorities realized that state involvement was necessary to ensure that historic preservation work was carried out effectively. Even today, Charleston's planning documents and ordinances dating from the 1920s represent an important source of reference for historic districts in U.S. cities. Chase (in: NTHP 1998, pp. 6) here contradicts Silver (1991), who sees the neighborhoods and private organizations as being instrumental in promoting the cause and success of the historic preservation movement.[143]

In October 1931, the first historic district of its kind in the U.S.A. was created in Charleston by the City Council. The Special Committee on Zoning initiated the work projects which, however, did not apply to the city as a whole. Although the city administration had much to do with the creation of the historic district, the population was also in part responsible for the measures introduced. As a result, Charleston is often cited as a ''pioneer of historic preservation'' (NTHP 1998). Charleston is one of the few cities to introduce historic preservation measures during the Great Depression (Cullingworth 1997, p. 114). The combination of private interests, urban planning, historic preservation regulations and objective preservation criteria greatly benefited Charleston (Chase in: NTHP 1998, p. 8). "The blending of planning and preservation goals was unique and a revolutionary concept for this time." (Chase in: NTHP 1998, p. 8).

At the same time, the Board of Architectural Review (BAR) and the Zoning Board of Adjustment were founded. At first, the committees existed merely as "negotiators"; in other words, they were to find a mutually acceptable compromise between investor interests and historic preservation concerns. Moreover, the city catered to the interests of historic preservation. As part of President Roosevelt's New Deal Program, it provided financial support to build municipal housing. Public funds were also made available to counter the damage caused by a hurricane that swept through Charleston in 1938. In 1939, the historic district project was broadened to take in the entire city. This enabled valuable buildings from the antebellum period to be preserved as part of a new, family-oriented municipal housing project initiated by the Housing Authority of Charleston. 1940 saw the setting-up of the first statistical inventory of Charleston's historical buildings. This showed Charleston to possess 1,168 significant historical buildings (this first statistical inventory was compiled by Helen Gardner MacCormack in 1940). This systematic listing was instrumental in the creation of the Charleston National Historic District.

---

[143] Gutschow (1978, p. 140) also demonstrates how effectively private groups of citizens defended themselves against marginalization and speculation.

After the Second World War, historic preservation in Charleston declined to some extent. Houses were damaged by further hurricanes and some, which could not be saved, had to be demolished. Towards the end of the 1950s, the Historic Charleston Foundation (HCF) began to revitalize entire neighborhoods by implementing projects based on aggressive purchase and resale strategies. Consequently, one of the first measures it was to take was the introduction of the urban historic preservation project known as the Revolving (loan) Fund (Silver 1991, p. 77). This was based on the idea that a small number of successfully preserved houses would attract buyers willing to undertake renovation work themselves. It favored buyers who expressed themselves willing to maintain and restore their newly acquired properties (see also Hofmeister 1977). This furthermore guaranteed that individuals had a personal interest in establishing and maintaining neighborhood structures. The 200 or so housing units that were restored by the HCF led to a gentrification process that excluded less wealthy social classes (Silver 1991, p. 77). In the same year (1959), the Historic District Zoning Ordinance was modified by the city authorities in such a way that the Board of Architectural Review could now act to prevent buildings constructed within the Charleston Old and Historic District prior to 1860 from being converted or demolished. Until then, no design guidelines had existed as yet within the Charleston historic district. The HCF could and was forced to act as it did, "..since Charleston lacked statuary power to engage in urban renewal until the 1970s .." (Silver 1991, p. 77).

During the 1960s, Charleston went into social and economic decline, making it imperative that new concepts be found to revitalize inner-city areas. As a result, some historic buildings were demolished in 1967 to make way for a city conference center. This contradictory decision clearly indicates the conflicting structures at work within the process of revitalization. Nonetheless, the potential profit to be made from tourists attracted to the city in the wake of historic preservation schemes was recognized, and Charleston Old and Historic District tripled in size as a result. In 1974, the Feiss-Wright-Anderson Survey and the Charleston Preservation Plan were created. The first exhaustive scientific and systematic inventory (since 1944, when an attempt was made to set up a list and failed) showed there to be 2,800 buildings worthy of listed status. The plan worked to preserve these listed buildings and develop a planning strategy that would last for several decades. The most significant guidelines included were the height regulations, introduced to ensure the preservation of the city's image, and a greater attention to the ratio of apartment units to housing units; in other words, fewer buildings should be divided into smaller apartments in order to combat the gentrification process.[144]

When Joseph P. Riley was appointed mayor in the 1970s, social acceptance of historic preservation in Charleston rose perceptibly. Riley considered it particularly important to heed the proposals of the Urban Design movement with regard to the integration of old and new structures. Fearing that Charleston would, like other cities, fall prey to suburbanization, Riley passed plans to build a modern conference center (the Charleston Center) in 1978. Many opponents to this plan feared that it would destroy the historical neighborhood structures and with them,

[144] Such units are described in American English as having been "condo-minimized".

the area's personality, as had so often been the case in Urban Renewal projects. Supporters, on the other hand, saw the Charleston Convention Center as a great opportunity to revitalize the city.

In 1988, African-American residents, who belonged to the lower-income group, successfully opposed plans to extend the Charleston Old and Historic District further. The City Councilor, Robert Ford, played a particularly prominent role, fearing that the proposed plan would lead to African Americans being pushed out of the areas by the white population (Silver 1991, p. 79).[145] In many southern cities, African Americans opposed plans to extend historic districts (Silver 1991, p. 79).

In the wake of Hurricane Hugo in 1989, which damaged 85% of all urban buildings to a greater or lesser extent, further discussions arose on the effectiveness of historic preservation measures in Charleston. The natural disaster led to a consortium being set up (Iv-Cha-2). This group helped property owners to keep economic losses to a minimum and assisted them in obtaining funding to carry out repairs. The city, however, was not prepared to override local laws in such matters and this resulted in financial hardship for many under-insured property owners. The number of cases brought before the BAR doubled from 600 to 1,200. With hindsight, we can conclude that many historical buildings were saved from being irreversibly converted only as a result of these strict laws.

In 1990, the Waterfront Park was set up by the city. This contains many park areas for recreational purposes, as well as offering spaces designed to house events and a restricted number of food stalls (Photograph 7.3). This form of harbor-area revitalization is a common concept within the United States and one which has been introduced with a varying degree of success. It was not until 1991 that the Visitor Reception and Transportation Center (VRTC) was opened in Charleston's Meeting Street on the edge of the historic district. The center combines historic preservation and the interests of the tourist industry at the very highest level. It is situated in a former train station dating from 1856, and its former function can still be seen from the building's facade. Tourists and residents alike can benefit from the transport on offer, the Down Town Area Shuttle (DASH) trolleys[146]. This reasonably priced and city-subsidized form of public transport comes with the additional advantage that visitors may park their cars at no extra charge at the Visitor Center. Thus, traffic within the historic districts is kept to a minimum, which is obviously in the interests of the residents (Iv-Cha-2).

---

[145] We are not necessarily dealing with gentrification here, as the district was not set up in order to enable a revaluation of the area.

[146] In this case, streetcars with rubber tires.

7.1.3 Current Planning Conflicts

7.1.3.1 Urban Marketing, Tourism and Historic Preservation

Charleston is currently working to create an urban atmosphere which applies, but is not just specific, to its historic district. A certain number of tourists will even be directed away from the historic area and its buildings. Thus, several paths have been set up alongside the Ashley River, boasting municipal features and vending stalls.

Charleston's reasons for developing alternative sites for tourists and neighborhood residents or the general population of the city and the SMSA region are of particular interest. Consequently, we find, on the one hand, urban marketing aimed at residents and visitors alike, whereas the population increases within tourist areas that are visited and used throughout the year. A purely historical atmosphere, which suits a museum but is ill adapted to communicate a vital, living "spirit", only allows the urban marketing industry to offer a limited, "one-day" experience.

As well as these permanent activities, a wide range of temporary attractions is on offer, such as the regular festivals that take place in Charleston. By creating the VRTC, the city is able to give in to neighborhood pressure, channel the incoming stream of tourists and manage the historic district areas efficiently.

Tourism in Charleston is a highly developed industry and one of its recent economic successes. Statistics show that tourism has enabled considerable changes to be made within the historic district.[147] However, it has now been realized that a significant conflict of interests exists between the city and the residential historic district. This can be seen within the various neighborhood organizations that exist in Charleston. According to residents, too many tourists pass through the district, causing disagreeable levels of noise and even instances of vandalism. Moreover, the residents believe their quality of life to be threatened (NIMBY phenomenon). As a result, there are no public restroom facilities or food trucks within the historic districts, in order to ensure that tourists' visits are kept short. These regulations were set up through the efforts of committed residents. Visitors' attention is drawn to the lack of facilities by leaflets and information that is distributed to them.

The large number of students living in the area poses a further problem. Charleston College has some 20,000 students, of whom a large proportion does not live on campus. On the one hand, the city is keen for a college to exist and produce qualified graduates, while, on the other hand, residents in the historic district are not interested in offering short-term leases. What is more, students are less likely to identify with the area in which they live, which necessarily reduces their individual sense of responsibility. Moreover, a constant tenant turnover significantly increases wear and tear to a building's structural fabric.

As well as the problems discussed so far, we should also mention the municipal housing unit project introduced in areas bordering on the historic district, which were funded by the Affordable Housing Program. In the 1960s, planners hoped to

[147] Mr. Sheehan (City of Charleston, Ombudsman) in Iv-Cha-5.

achieve a social mix within the historic district, thereby integrating various social classes into the community. Although these apartments were built before the Civil Rights legislation of 1964/65, it was hoped that ethnic integration would take place to a certain extent. The measures taken should combat segregationalist tendencies.

The neighborhood organizations, which have recently pooled their interests and joined the Charleston Consortium, work to increase public awareness of problems identified by both groups (for example, crime, drugs, vandalism and noise pollution). Within this association, members work alongside police and have set up their own patrol system (Iv-Cha-2).

7.1.3.2 Zoning and Construction Planning

A further significant problem in the Charleston historic district is the coordination and execution of construction plans. This is particularly problematic when the plans propose changes be made to a building, including interference with the original historical structures in order to convert former factories, retail units and shops into apartments. The conversion of larger buildings into smaller housing units that attract "gentrifiers" is a common form of intervention. However, constructional change also reflects social and structural change (Iv-Cha-2). Some residents are extremely skeptical of this tendency and see it as a threat to the advantages of life in the historic district[148] (Iv-Cha-1, Iv-Cha-2).

It is interesting, at this point, to mention the example of the People's House, which financial investors wanted to convert into apartment units (Photograph 7.4). Several city projects have also attempted to attract new industries to the city center. Public authorities hoped that this would help their marketing strategies and improve Charleston's locational quality. The desire to convert is thus undoubtedly justifiable, yet such schemes would drive older industries away from the center (Iv-Cha-1). Cullingworth cites several cases where historic district residents also found the construction of skyscrapers in neighboring areas threatening (1997, p. 117).

Since the 1990s, a general trend can be identified in Charleston that accords greater importance to historic preservation within the historic district. This is particularly evident in rapidly growing districts (Chase in: NTHP 1998, p. 14). Historic preservation laws in such cases should, of course, ideally be binding.

Even today, no legally binding historic district zoning ordinance exists in Charleston city, which does much to complicate negotiations with public authorities. Birch and Roby (1984, p. 199) nevertheless describe zoning schemes in Charleston as exemplary.

[148] NIMBY phenomenon.

7.1.4 Charleston: The Planning Type

From a planning perspective, Charleston is a very "mixed bag". Despite the local authorities' comparatively committed involvement in historic preservation after 1931, through the implementation of Urban Design and planning schemes, there are still many gaps to fill within existing measures for historic preservation. Chase (in: NTHP 1998, p. 14), comments that preservation measures were successful not as a result of buildings being grouped together or urban (administrative) structures, but as a result of citizen involvement. Interestingly enough, Charleston is home to one of the few U.S. historic districts that have no design guidelines. This is due, above all, to the long-standing historical status of the district and its administrative structures.

Although the Charleston Old and Historic District was set up mainly thanks to well-structured planning schemes, the private influence of the HCF should be mentioned here. It was a combination of the former and the latter that enabled the historic district to take the form we know today. The main problems in Charleston's historic district are mainly related to the tourism industry, which leads to a conflict of interest between the aims of the industry and the residents' wishes (Iv-Cha-4). Zoning is a further problem in Charleston. Attempts are currently being made to allow new industries to move to the historic district and relax construction legislation to permit the conversion of old buildings into small residential units. Some neighborhood organizations have already made their opposition to these schemes clear (Iv-Cha-2). It must be noted, however, that Charleston is one of the few cities that actively adopts historic preservation measures to benefit economically disadvantaged social groups (Silver 1991, p. 80).

Finally, we must mention the fact that Charleston competes fiercely with Savannah to attract tourists to the city. We should, however, note that it is preferable to see tourism as a common cause affecting both Charleston and Savannah than as a divisive factor that causes them to compete on a regional level.

The examination of Charleston demonstrates that historical buildings have been protected by means of historic preservation schemes despite the existence of certain limitations (see also: Fitch 1995, p. xii). Compared to Society Hill and Beacon Hill, Charleston city's planning activities have come to the forefront in recent times. The conflict of interests that arises due to the tourist industry's activities within the city has yet to be resolved by residents and officials. The historic district in Charleston has been gentrified to a certain extent, yet has simultaneously worked towards heterogeneity; in other words, various social and ethnic groups have also been integrated. Initiatives similar to the Affordable Housing Projects in Society Hill or Beacon Hill were successfully opposed in the 1970s by the neighborhood organizations.

Photograph 7.1. Charleston Old and Historic District with the prominent colonial-style houses in the residential section of the historic district.

Photograph 7.2. Charleston Old and Historic District in the commercial district. Colors of the outside facades and individual designs must be officially approved.

Photograph 7.3. Waterfront Park in the southeast of the Charleston Old and Historic District was set up just a few years ago. It is intended to serve as an "urban recreational area" for residents, tourists and workers.

Photograph 7.4. The "People's House" (the largest building to the left of the picture) is currently an object of controversy because the former office building is proposed for conversion into apartments. Critics fear the historic district will be utilized in an excessively one-sided way. They support mixed use.

7.2 Savannah

7.2.1 Savannah's Urban History

Native Americans settled early on in the area today known as Savannah. King George II entrusted the area to some of his trusted followers and decreed that they work to strengthen the colonies even further and ensure that they became a further source of wealth for England. It was General James Oglethorpe and his philanthropic supportings who wanted to create a safe haven for all Protestant groups who risked persecution in the Old World. Moreover, the colony of Georgia[149] was expected to protect Carolina from the Spanish forces in Florida.

[149] The colony of Georgia became the state of Georgia after the American Revolution.

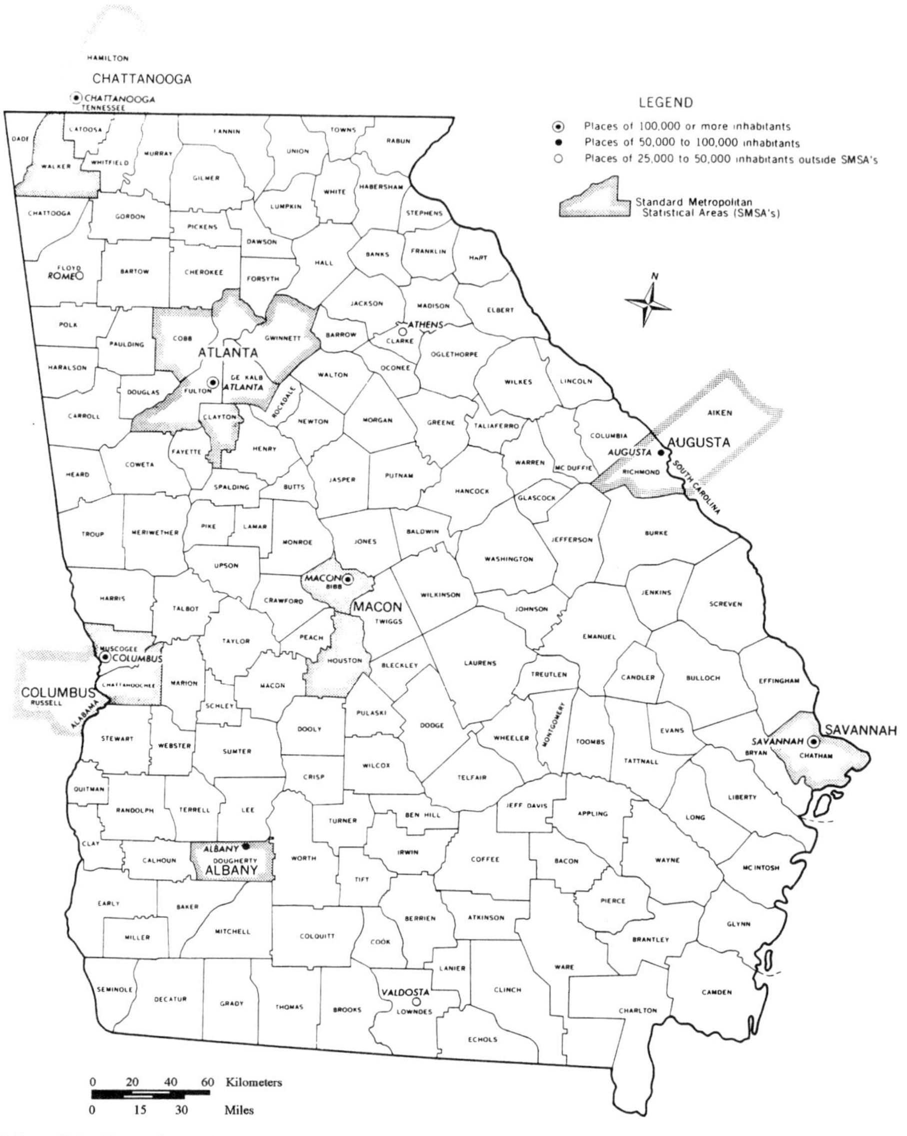

Map 7.5. Georgia and the Savannah SMSA (Source: Andriot 1979, p. 156)

The harbor city of Savannah on the Savannah River was founded in 1733 by English colonists (Map 7.6). By the same time, plans were proposed for Savannah, drawn up by the urban planner and general, James Oglethorpe, in England.

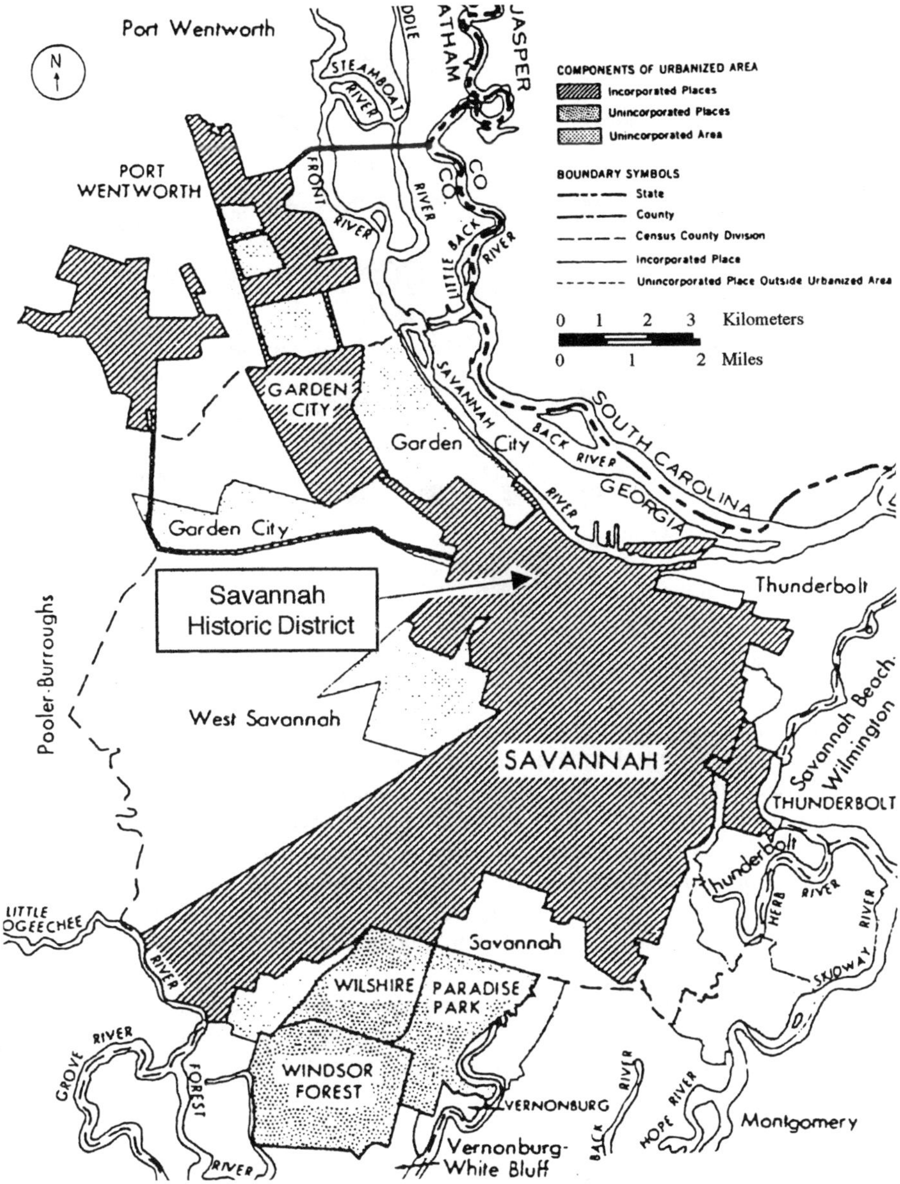

Map 7.6. City of Savannah (Source: Andriot 1979, p. 167; modified)

A: Franklin Square
B: New City Market
C: Ellis Square
D: Telfair Square
E: Orleans Square
F: Pulaski Square
G: Chatham Square
H: Johnson Square
I: Wright Square
J: Chippewa Square
K: Madison Square
L: Monterey Square
M: Forsyth Park

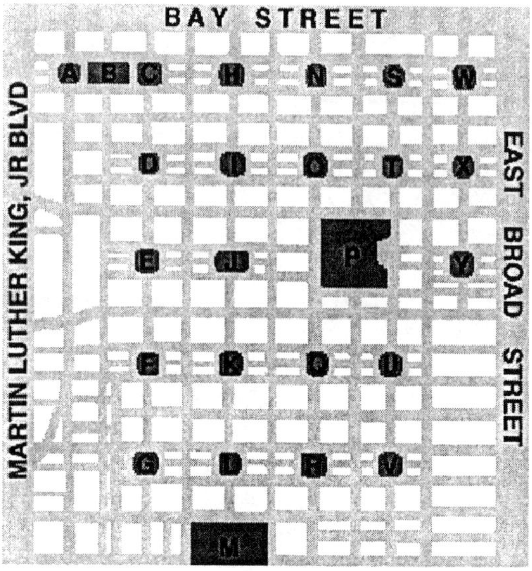

N: Reynolds Square
O: Oglethorpe Square
P: Colonial Park Cemetery
Q: Lafayette Square
R: Calhoun Square
S: Warren Square
T: Columbia Square
U: Troup Square
V: Whitefield Square
W: Washington Square
X: Greene Square
Y: Crawford Square

Map 7.7. Savannah's sophisticated grid (Source: www.savannahnow.com, 8th April, 2000)

The plan proposes a city with streets laid out in a sophisticated grid of wards, each comprising a central square flanked by public buildings ("trust lots") to the east and west, and lager residential lots to the north and the south (Map 7.7). The squares in the east and west are spacious enough to accommodate the construction of public buildings such as schools or libraries. Private plots of land were available in the north and south, designed for the construction of housing (Lyon in: NTHP 1998, p. 37).[150] Savannah may be distinguished from other cities dating from the early settlement period due to the city's particularly interesting grid (Hofmeister 1977, p. 138).
Once this proposal was passed and implemented, Savannah was to become one of the first planned cities in the United States. The plan became famous, and with it, Savannah. To this day, urban planners and architects travel to the city to study its structural layout.

[150] The first church to be built in one of these squares was the Christ Episcopal Church (at that time still the Anglican Church), which has since become the "Mother Church of Georgia".

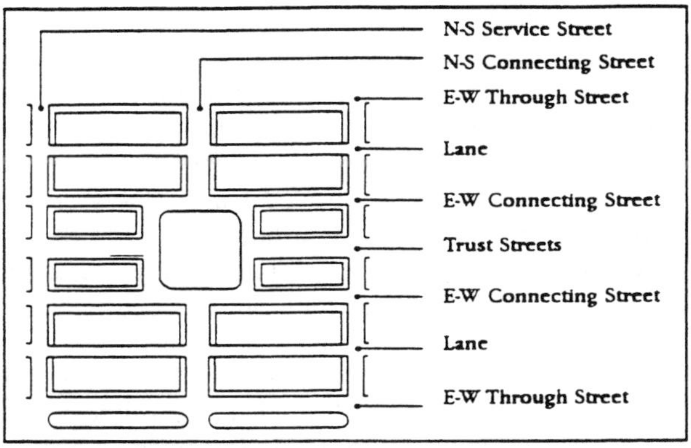

a) Squares and Connecting Streets

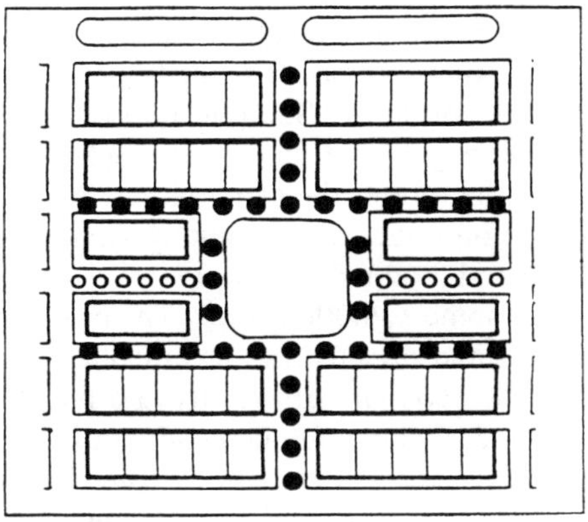

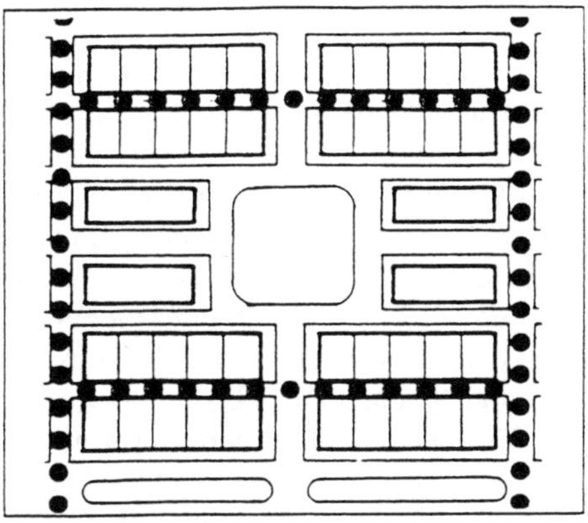

b) Neighborhood Streets c) Service Streets and Lanes

Map 7.8 a-c. Square (detail) (Source: City of Savannah 1997a, pp. 2–3)

In 1754, Savannah became the administrative capital of colonial Georgia. Savannah was to become an important port, in particular prior to the American Civil War. Mainly 19th century buildings, emblematic of a flourishing economy, have been preserved within the city to this day. Among the first Europeans to settle in historic Savannah between 1733 and 1776 were many Jews, who came to the city from Portugal to flee the Spanish Inquisition. The well-known and established Mickve Israel (Old Jewish Congregation in the South) evolved from this group. Protestant groups from Salzburg, Scotland, Holland and Ireland were to follow, some of them facing persecution in their homelands. Significantly, there was no slavery in Savannah until 1750, it was forbidden to sell rum until the 1750s and there were no lawyers or Catholics in Savannah until 1755. Yet it was soon realized that the good intentions behind such rules (law-abidingness, alcohol prohibition) could not prevent the decline of Savannah society and they were abolished.

Until the mid-19th century, Savannah's wealth derived from its thriving cotton trade. To this day, people speak of the cotton belt that ran through the region. The silk farms to the north of the city and the cotton plantations and transportation were not cost-effective. Shortly after this became clear, a third of the population consisted of slaves. Thus, slave dealing with Africa and the selling of slave workers made some members of Savannah society very wealthy.

Antebellum Savannah was thus a wealthy city. Nevertheless, it was affected on many occasions (1796 and 1820) by major natural disasters, which caused permanent damage on some buildings and destroyed many others. Yet Savannah succeeded time and time again in overcoming a hurricane and other (natural) disasters, such as a yellow fever epidemic, and on every occasion managed to reconstruct itself in the mould of its rich and glorious early years. However, Savannah's economic foundation, the cotton trade, whose success was due mainly to the work of the slaves, was soon to face a new threat. In 1850, political tensions between the North and the South grew and reached an all-time high in 1960. As its economy was so dependent on the slave trade, Savannah joined the Confederate side during the American Civil War. In 1864, many citizens of Savannah left the city and thus buildings did not suffer any great damage as a result of the war.

The Reconstruction period began. For some time, it was difficult to persuade slaves who had been liberated and chosen to remain in the South to return to the farms as part of a free and paid workforce. The gulf that separated blacks and whites began and became wider and wider. At the turn of the century, new industries came to the area. The once flourishing cotton trade was eventually swamped by the economic superposition of the chemical and food industries.

Today (1998), Savannah has a population of some 130,000 people. The city has attracted many academics thanks to its colleges and famous Savannah College of Arts and Science (SCAD) founded in 1979. The economy of the city and the region as a whole has grown as a result of the food, timber and chemical industries. High-tech manufactures, especially Gulfstream Aerospace, attracts many engeneers and other specialists. Moreover, a crude oil refinery is situated in the area, and the city also benefits greatly from the tourist industry. In 1999, Savannah's SMSA catered for around 6 million tourists and made a total of 634 million U.S. dollars (Iv-Sav-2). The fact that the U.S. population can now expect to live longer is seen as a positive phenomenon by the Tourist Bureau, a factor significantly boosting Savannah's tourism industry. People have more leisure time that they can spend in Savannah's historic district.

7.2.2 Beginnings of the Historic Preservation Movement

Savannah's fame as a well-preserved and attractive city can be traced back to several women pioneers in the field of historic preservation. In 1945, these women were eventually to become the trustees and patronesses of several individual historic buildings. In some cases, these women were able to persuade companies to restore buildings housing business representatives. Financial and economic successes did much to convince skeptics of the value of historic preservation.

In 1955, these women, led by Anna C. Hunter, were able to save Davenport House from decay and demolition. A short time before this, the Historic Savannah Foundation (HSF) was founded and continues to exist to this day, using donations to preserve and restore buildings. Until 1968, the HSF successfully implemented the Revolving Fund System. The state did little to introduce guidelines and measures. However, it did play a more active role within the historic district, as will be discussed below in more detail.

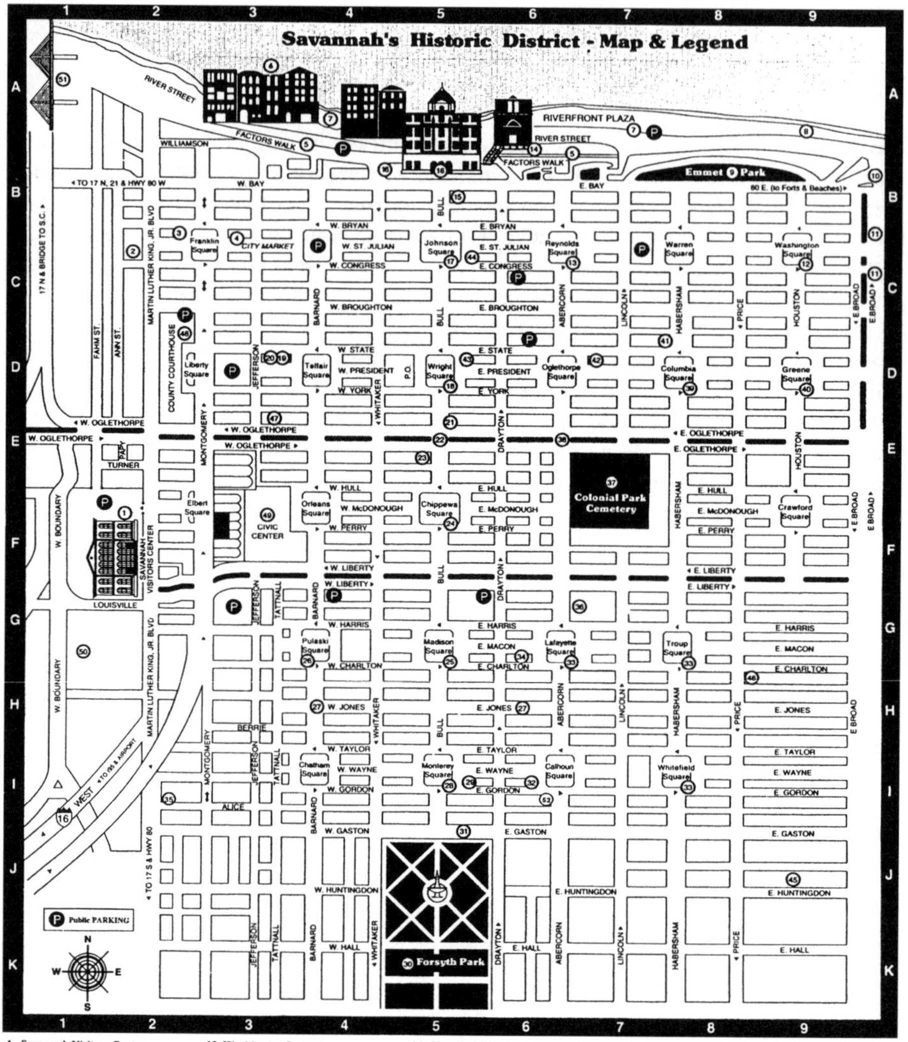

1. Savannah Visitors Center
2. Scarbrough House & Ships of the Sea Museum
3. First African Baptist Church
4. City Market
5. Factors Walk
6. GA's oldest stone & masonry bldg.
7. Riverfront Plaza / River Street
8. Waving Girl statue
9. Emmet Park
10. Fort Wayne
11. Trustee's Garden Village
12. Washington Square
13. Reynolds Square
14. Cotton Exchange
15. U.S. Customs House
16. City Hall
17. Johnson Square
18. Wright Square
19. Telfair Mansion & Museum
20. Trinity Methodist Church
21. Juliette Gordon Low Birthplace
22. Oglethorpe Avenue
23. Independent Presbyterian Church
24. Chippewa Square
25. Madison Square
26. Pulaski Square
27. Jones Street
28. Monterey Square
29. Temple Mickve Israel
30. Forsyth Park
31. Gaston Street
32. Wesley Monumental United Methodist Church
33. Whitefield, Troup & Lafayette Squares
34. Andrew Low House
35. R.M. Gilbert Civil Rights Museum
36. Cathedral of St. John the Baptist
37. Colonial Park Cemetery
38. Oglethorpe Avenue
39. Columbia Square
40. Greene Square
41. Davenport House
42. Owens - Thomas House
43. Lutheran Church of the Ascension
44. Christ Episcopal Church
45. King - Tisdell Foundation & Cottage
46. Beach Institute
47. Savannah Area Chamber of Commerce
48. Chatham County Courthouse
49. Savannah Civic Center
50. Historic Railroads Shops
51. The Great Savannah Bridge
52. "The Book" Gift Shop

Map 7.9. Savannah's Historic District (Source: Savannah Visitors Bureau 1999)

7.2.2.1 In the Beginning..: Savannah's Historic District

The Georgia Historical Society (GHS), founded in 1839, had its headquarters in Savannah. This is the longest-standing cultural institution in Georgia and one of the oldest in the country. In terms of the tourism industry and the development of its historic district, Savannah has long been overshadowed by Charleston; yet the city has its own brand of charm that sets it apart from Charleston. We might even compare Savannah with New Orleans, though this once again serves mainly to distinguish Savannah's own individual character. Savannah is a leafy and idyllic city, thanks to the large number of trees and striking areas of grassland to be found there. The harbor area is competing with Charleston. Thus, the city is trying to attract as much attention and investment as Charleston receives. Today, the Savannah Historic District represent the largest historic district in the U.S.A. The following analysis will, however, focus exclusively on the Savannah Historic District, as the districts are administered according to different criteria and subject to different regulations.

Savannah's historic structures were left more or less intact throughout the Civil War period and post World Was II, when the International Style[151] was modern in the United States. However, the city was unable to escape several irreversible construction plans from being implemented in the 1950s (also Hofmeister 1977, p. 138). In order to compete with the shopping centers being set up in the suburbs, the old Savannah City Market was demolished to make way for a multi-story parking garage. This garage is still known today as the City Market (Lyon in: NTHP 1998, p. 38).[152] Many buildings in downtown Savannah were abandoned, as residents moved to houses lying south of the old town. About a third of the historical structures were destroyed between 1930 and 1950. Plans existed to convert all of the unique squares, opening them up to traffic. Yet it is precisely this system of squares that makes Savannah so unique. Already, some squares had been converted to allow fire trucks and large vehicles to pass through (Lyon in: NTHP 1998, p. 38). Three squares on Montgomery Street was converted in 1935 and traffic circulation greatly improved (Lyon in: NTHP 1998, p. 38). A similar project was initiated near the Old City Market, replacing the square with a multi-story parking garage built of gray concrete. Some 18th century houses were demolished, to be replaced with a number of modern constructions. In doing this, the city was imitating the contemporary architectural trends evident in the suburbs, whose population was growing steadily at that time as a result of the suburbanization phenomenon.

A group of seven women preservationists formed the Historic Savannah Foundation (HSF) in 1954 in answer to the widespread demolition and defacing of historical structures that was taking place at the time (Birch and Roby 1984,

---

[151] See Glossary for an explanation of this term.

[152] This garage is due to be demolished in a few years and the square transformed into a greenspace.

p. 199).[153] At first, the HSF was made up of voluntary workers, though paid full-time workers were later employed. Initially, the HSF concentrated on preserving isolated buildings, developing into a professional organization that eventually worked to set up the Revolving Fund Program (Costonis 1974). Houses were sold to buyers willing to restore the property. This method had the advantage that whole neighborhoods could be protected, buyers could be selected up to a certain point, and house prices could be regulated. Savannah banks offered credit to potential buyers, though the system favored the private sector above the public sector (Silver 1991, p. 78). In the absence of zoning ordinances and restrictive design guidelines, buyers could be attracted and persuaded to carry out historic preservation work that would fulfill the ideals and concepts of the HSF (Lyon in: NTHP 1998, p. 39). Furthermore, the HSF was able to broaden its scope and thereby qualify for other funds, from the Urban Renewal programs and HUD projects, to support their own preservation work. As well as the transfer of real estate, they worked to improve public buildings, parks and trees. In the late 1980s, a square that had just been converted and opened up to traffic (Franklin Square) was reclaimed and restored to its original condition.

The women who founded the HSF at that time wanted to win private support to promote public preservation programs. Their aim was to establish the idea that it was more rewarding to invest a million dollars in a city's historical structures than to develop and build new structures that clashed with the city's architectural and cultural heritage.

The public awareness that the HSF succeeded in developing through its work attracted the attention of the renowned banker Leopold Adler II. Their efforts resulted in the creation of an historic preservation strategy unique to Savannah, based on the following three criteria (Birch and Roby 1984, pp. 199):

1. A statistical record was to be kept in order to achieve an overview of existing historical features,
2. a campaign should be organized for the creation of an historic district through universal construction plan guidelines,
3. a bank account should be opened in which donations could be deposited to help historic preservation in Savannah, fund long-term and larger projects and to give historic preservation in Savannah a more or less official status.

From 1962 to 1968, a complete list of the statistical and architectural features of all existing historical structures was compiled. In 1966, the city passed legislation designating the Savannah Historic District, and Savannah was named a National Historic District. In 1968, the historical inventory of the National Historic District was published, an important step towards further developing public awareness

[153] It was the proposed demolition of Davenport House (built in 1821) that motivated these women to take action. With funding from donations they were able to preserve the house. Yet the demolition fever did not die down. In 1960, Armstrong College wanted to demolish six historical buildings in Bull Street and replace them with new buildings. After further consideration, it was decided that they should merely be converted.

(HSF 1979).[154] 1,100 historical buildings were identified as contributing resources within the urban context. As yet, no attempt had been made to create a statistical inventory of other buildings that might be considered non-contributing resources.[155]

For the first time, in 1965, the HSF produced a publication intended to promote tourism, entitled "Savannah, Georgia as a Travel Destination" (HSF 1965). The next step was then to inform and motivate the public to become actively involved. The National Preservation Conference was held in Savannah in 1968 as a direct result of the HSF's initiative. Until then, however, planning components had not been incorporated into historic preservation schemes (Silver 1991, p. 78). The regional Chatham County Savannah Metropolitan Planning Authority, which focused on Savannah's SMSA, finally developed a preservation plan for the city in 1969.[156] This included 16 design guidelines, to be applied to buildings and structures within the historic district. A few years later, the department of finance produced a further plan that concentrated on the economic development of the historical downtown (Savannah Housing Authority 1969). It was several years, however, before these guidelines and planning proposals were legally implemented (Lyon in: NTHP 1998, p. 40).

7.2.2.2 Extension to the Savannah Victorian District and Surrounding Area

After 1970, historic preservation won widespread public and official acceptance. The pioneering work undertaken by the HSF from 1955 to 1968 made it possible to lay the basis for structures that are still in place today. Nowadays, local and national governments play a much greater role than was the case in the early years of the historic preservation movement. Several organizations have been founded in Savannah and Georgia, whose exemplary nature greatly influences national programs of historic preservation. When the National Historic District was established, the local government was obliged to develop a local historic district ordinance. The administrative structures and inventories created in the early 1970s served to identify significant historical buildings in Savannah. The Victorian District[157] was included in the National Register in 1973. Even the relatively small

---

[154] In the second edition, which appeared in 1979, the Victorian District was also included.

[155] The department of Architectural History at SCAD is actually conducting a research of all buildings and structures (historic and otherwice) in Savannah historic district. This is part of the Virtual Historic Savannah Project (directed by Prof. R. Williams). www.scad.edu/virtualsavannah/

[156] Hofmeister (1977, p. 149) makes a similar observation. Existing historical societies had a large degree of social control.

[157] Above all, it must be noted that in no other U.S. city has a district with so large a number of Victorian buildings been preserved. Plans are currently being developed by neighborhood organizations to protect Boston South End, while in Philadelphia and West Philadelphia the Philadelphia Historical Commission is looking into the feasibility of implementing a similar scheme.

Central Georgia Railroad District to the west of the city was included on the request of the HAER in 1975, and declared a National Historic District in 1978 (Lyon in: NTHP 1998, p. 41). The old station site is host to the impressive Savannah Visitor's Center and Savannah History Museum.

The manner in which historic preservation was carried out in Savannah did little to benefit the African American community and directly contradicted their grass roots philosophy, leading, in 1977, to the foundation of the activist group, the Association for the Study of Afro-American Life and History (ASAALH) by the African Americans. 51.3% of Savannah's population is made up of African Americans (U.S. Census 1990). This activist group saw itself as providing an alternative to the ".. white dominated landmark initiative" (Silver 1991, pp. 79). It becomes clear, in fact, that historic preservation proved prohibitively expensive for less wealthy social groups. Structural data gathered by U.S. cities often points to a discrepancy between results obtained in the SMSAs and national averages. In 1990, 22.6% of the population in Savannah lived below the poverty line; in Georgia as a whole, this figure was reduced to 11.5%, whereas nationally, only 13.5% were considered to live in poverty (U.S. Census 1990). In 1989, the average wage was $22,102 within the city of Savannah, whereas in Georgia it was $29,021. Still, the less wealthy Victorian District was extended by about 25% in 1982. Today, it contains 425 historical buildings (contributing buildings) and 25 more recent buildings (non-contributing buildings).[158]

It was not until the late 1980s and early 1990s that efforts began to compile an inventory listing the historical resources left by the history of African Americans in Savannah. These include, for example, the First Bryan Church or the Laurel Grove Cemetery and some smaller historic districts bordering on the main historic district that had been established in 1974. Today, there are plans to incorporate the listed buildings into the National Register, although it still remains for them to be included in national preservation schemes. Westley W. Law proposed that a Negro Heritage Trail be set up, taking into account particular locations and features reflecting the history of the African Americans.

Savannah introduced innovative measures in a further program carried out in the Victorian District. Prior to 1955, the structural fabric of the buildings had been in an extremely poor condition and the residents did not have the financial means to carry out the necessary renovation work. Special individual and national programs were introduced to provide aid for such urgent cases. From 1974 to 1989, 500 apartments were built in the "hopeless" slum area. In 1980, the HSF had created housing units to cater for middle class residents. Heated discussions took place to decide whether rehabilitation could be considered a means of improving residential structures (Lyon in: NTHP, p. 42). After a meeting held in Savannah in 1994, the Advisory Council on Historic Preservation (ACHP) had investigated these factors and introduced new guidelines for the Affordable Housing Programs (Lyon in: NTHP, p. 42). The subject was discussed increasingly frequently, as it became clear that wealthier districts could set up individual historic preservation projects to renovate the area and in turn be transformed into protected historic

[158] Source: Georgia Department of Natural Resources, Historic Preservation Division.

districts. Historic preservation work was carried out within the Savannah Victorian District without residents having to be (Sander and Wilkinson 1989, p. 56). The historic district would be established by a residential community having common interests, with the aim of preserving buildings and, at the same time, ensuring the stability of the neighborhood organizations (ibid. pp. 56).

The Savannah College of Art and Design (SCAD) was founded in 1979. This involved the renovation of a total of 47 historic buildings within the CBD. Several motels were converted into halls of residence and student accommodation. This ensured that the area was revitalized and security increased, while also achieving a mix of age groups and social classes within the Savannah CBD. Moreover, this made it possible to develop a cultural program aimed at attracting tourists to the area. At the same time, more courses were introduced and plans exist to introduce M.A.

In the years that followed, the city set up two more Urban Renewal Projects and began extensive restoration work to the harbor area. Many restaurants, jazz clubs and shops have been set up there as a result. Although Savannah is looking to develop its own individual image, the docklands are still referred to as Baby New Orleans (Savannah Visitors Bureau 1999).

7.2.3 Current Planning Policies

Savannah's planning procedures, which remained ineffective until the 1970s and which Silver (1991, p. 78) criticizes in his work, are now being improved with the incorporation of several measures proposed by the city. The urban population and city administration now believe that tourism can greatly benefit the historic district (Iv-Sav-1). One problem that persists, however, is that no corresponding plans exist to convert and expand the relatively large number of museums. If museums are not large enough to welcome visitors, this could lead indirectly to a loss of vitality in the historic districts, as the daytime population continues to increase.

An important factor in the stabilization of downtown areas is undoubtedly the establishment of a group of permanent residents. This is the main objective of programs run by the SCAD. The city has recently agreed to lend its support to these programs. It is particularly concerned that residential projects should include all ethnic groups in their preservation plans. Here, all basic social considerations included in Savannah's education programs play a vital role. The construction of the Massie Heritage Center as part of the 1975 Heritage Program has done much to win acceptance and develop awareness in the context of heritage awareness.

In the longer term, Savannah is looking to create its own individual image in order to prevent the city being compared with Charleston (because of the historic district), New Orleans (the docklands area) and Atlanta (shopping facilities). The city's nomination by the International Council of Monuments and Sites (ICOMOS) will certainly enable it to stabilize its image and develop its tourist industry further. Unfortunately, this nomnation has not being actively persued during the last few years. For the time being, procedures have reached a stalemate, due to the fact that the city and its residents only want to see its structural layout nominated, and not its buildings (Iv-Wash-1).

As a result of the many natural disaster that have affected the area in the past, the state of Georgia has created a multilateral disaster prevention system (see Chapter 4) as part of the Preservation Planning Program that is applicable to both the Savannah Historic District and the Victorian District. This provides for the following:

1. The safeguarding of public and financial resources for the purposes of historic preservation;
2. Preventive measures to protect and safeguard certain areas;
3. Provisions for the implementation and dismantling of a network;
4. Work with the public and call for citizen cooperation (in the area of heritage education programs);
5. The gathering of information.

7.2.4 Savannah: The Revolving Fund Type

Savannah, like all cities, has experienced problems that were either already present or developed during the 1950s. These problems included the growing tendency towards massive suburbanization, the population's increasing mobility, the call for more park areas in the city and the departure of the manufacturing industry as well as other traditional industries. These problems were to some extent approached by introducing historic preservation programs; in other instances, urban programs and plans directly opposed the aims of historic preservation. Large-scale projects made it necessary to demolish historical structures. Savannah's original city plan did not foresee the need to create additional parking facilities.

Had the pioneering work of the HSF never taken place between 1955 and 1968, Savannah would probably not be so well preserved today. By 1968, the HSF's Revolving Fund Program had led to the restoration of around 130 houses (Hofmeister 1977, pp. 146). Private initiatives and involvement have succeeded in increasing interest in historic preservation within the city of Savannah and have set the wheels of change in motion. The local government could only be persuaded to act once it had realized the positive effects that preservation schemes could have. Even today, additional guidelines are introduced and historical or architectural studies carried out thanks to private funding. It was not until public and private associations cooperated (through tax benefits) that it was possible to initiate a far-reaching scheme to preserve the historic district.

Tall buildings with a higher return on capital contradict the concept that the historic skyline should be preserved. It is difficult to judge what economic effect (improvement in image, increase in tourist numbers) the historic skyline may have. In any case, this is a judgment that can only be made by the public authorities and not the investors themselves, as the effects will only become clear with the passage of time. For this reason, the city of Savannah ordered that a study be carried out in 1990 to include height recommendations setting down height restrictions for all areas of the city.

In Savannah, it is clear, above all, that the organic development of the neighborhood organizations and the absence of large Urban Renewal Projects did much to enable the preservation of historical structures (Fitch 1995, p. 37). Neighborhood preservation measures showed how it was possible to minimize the scale of relocation necessitated by historic preservation schemes (Silver 1991, p. 78). It has furthermore been proved that state participation is not necessarily required in order to preserve historic structures. In particular, historic preservation work was carried out in the Victorian District without alterations to existing neighborhood structures, relocation of residents or other changes becoming necessary (Sander and Wilkinson 1989, p. 56). This project mainly involved the participation of organizations with common interests at heart and followed the aims of the Affordable Housing Program. Although Savannah's structure, in terms of the evolution of its historic preservation schemes, is similar to structures evident in other cities, its projects and planning measures were less strictly organized. Savannah ".. underscored the diversity of the planning/preservation heritage in the South.." (Silver 1991, p. 78). Whereas both Charleston and Savannah have implemented the Revolving Fund system through private organizations, the government in Charleston was quicker to intervene in and oversee urban projects.

In recent times, some structural improvements have taken place as a result of historic preservation: tourists can observe the positive influence of the Savannah College of Art and Design (SCAD) on the downtown area: the presence of students during the daytime means that the streets are not merely peopled by tourists in the hours of daylight. The programs are being broadened to encourage this phenomenon. The revitalized and deliberately multi-functional harbor area can also be judged a success. The number of tourists, as well as the large proportion of local visitors who return to the area, serve to confirm this judgment.

Photograph 7.5. An image typical of the the Savannah Historic District. In the foreground an historic square from the original 1733 city plan.

Photograph 7.6. The old City Market, which today is the New City Market, has been converted into a garage for residents, tourists and employees in the historic district. The green awning is a visual reminder of the historic market.

Photograph 7.7. The harbor front as it has been revitalized under the Urban Renewal Program. The upper stories now contain professional offices housing architects, city planners and lawyers.

Photograph 7.8. A drive-through teller bank situated in the middle of the Savannah Historic District. The automobile still dominates here and there the historic district. Nevertheless, Savannah can be called a pedestrian-friendly city, the urban plan forces cars to slow dorn and navigate their ways around the squares, while pedestrians can move directly through the squares.

8 Summary of Findings

All four historic districts in the cities examined in this study play an essential role in U.S. urban history. This history[159], planned with urban political interests in mind and stage-managed from a social perspective, nevertheless clearly indicates a number of paths of choice, classified in this study under the four categories of preservation (Urban Renewal Type in Philadelphia, Neighborhood Type in Boston, Planning Type in Charleston, Revolving Fund Type in Savannah).[160]

In the area of practice-oriented urban planning, many experts have criticized the means adopted to assess success. It is fairly easy to investigate and evaluate the status quo of a country or city, whereas measures taken to assess success must also take into account past and present. In the combined fields of urban planning and historic preservation, many factors discussed in the analytical part of this study come into play, so that success must be measured according to different scales of evaluation. On the one hand, we have the structural/functional level, while on the other, we have the socio-historic level of evaluation. Both scales of evaluation are significantly different and, in the process of measuring success, can lead to significantly different results. The following chapter will, moreover, seek to differentiate between the levels of evaluation to be found on a national level (within the United States), as well as on an international level (a comparison will be made with Germany).

8.1 Urban Historic Preservation in the United States: A National Comparison

Historic preservation is today widely accepted by society. Methods are planned and financial support provided through private and public institutions, as has been demonstrated in the analytical part of this work. Hodder (1999, p. 439) confirms that there are to date more than 1.800 Historic District Commissions and Boards of Architectural Review (BAR) in the United States.

For many years no official structural guidelines or national program to aid regional development existed in the United States. This state of affairs to a large extent gave rise to the wild growth of architectural development evident in the

---

[159] We could also use the term "heritage" in this context.

[160] It is useful here to refer to the tables providing a more complete comparative evaluation. These can be found in Appendix B, Tables B.1 to B.9.

cities. Many buildings did not meet with necessary structural regulations regarding fire safety and hygiene. Uncontrollable architectural growth, suburbanization, urban economic decline and a generally unfavorable attitude towards the cities meant that some citizens came to recognize the city's shortcomings and its deficient planning system. It is only natural that regional planning and the question as to who one's neighbor actually is should personally affect the individual to some extent. Particularly since the 1970s, individual citizens have often participated actively in the shaping of their own environment and acted on their own initiative. In this way, historic preservation represents a means of achieving a homogenous neighborhood. The right to participate, open-ended procedures, changes and renewal work are made available to the public by means of public hearings on plans for development. It is remarkable how emotional and personal Americans' reaction can be to a few small changes. In the United States, planning and construction concepts fall very much within the public domain and are decided by local authorities. A small-scale or large-scale concept of regional organization and, by extension, of historic preservation, does not exist, or, at the very most, exists in a very rudimentary form.

The four historic districts in the cities of Philadelphia, Boston, Charleston and Savannah can be seen in many ways to have similar structures and organizational features, and their individual problems and potential can well be compared. In almost all cities, the beginnings of the historic preservation movement can be traced back to the idea that individual people should be responsible for individual objects.

In the 1950s and 1960s, all cities had to combat similar problems. These consist mainly of:

1. The decline of local industry and the ensuing lack of employment and economic stagnation
2. Suburbanization
3. The general attitude about urban life of the population
4. Definitions of history and culture (the development of heritage awareness)
5. Lack of parking facilities
6. The need for specific structures within the city, such as the construction of a highway or a conference center.

It has been proven that the individual dynamics of historic preservation in the different cities have developed as a result of these problems.

It is remarkable that significantly different methods have developed in each of the four cities to deal with historic preservation and establish historic districts. To this day, preservation work, such as organizational work and financial subsidization, is undertaken in very different ways in the four cities. In this way, we can clearly distinguish between the extent and focus of historic preservation within the four cities (see also tables in Appendix B.3 and B.4).

The historic district examined in this study are, as has been shown, the areas of a city most frequently visited by tourists. In many cases, areas and neighborhoods bordering on an historic district are still in a state of decline.[161] To judge the situation from an international and a national perspective, it remains to be seen whether these factors can be separated, or whether, in fact, they must be considered as a whole. Comparisons should thus be restricted to the basic factors that play a role within the individual countries.

The approved programs of historic preservation in the United States constantly come up against their own limitations. No universal system of classification exists to evaluate and substantiate systematically and nationally the continued existence of American buildings worthy of preservation. Due to the relatively poor documentation available, it is, in part, impossible to ascertain whether some buildings at one time declared to be of historical significance still exist and in what condition they may be. It is up to individual local governments to implement an efficient local system of historic preservation according to their own evaluations and criteria. Statistical information is consequently clearly the concern of regional organizations.

One of the fundamental disadvantages inherent in the practice of historic preservation within the four historic districts discussed here is the fact that permanently disadvantaged areas are also treated as second-class regions by those in charge of implementing programs of historic preservation. In Savannah and Boston, an effective solution was found to this problem by introducing the Affordable Housing projects (Datel 1985, p. 141). An examination of the area's aesthetic and architectural features allowed the social and historical character of the neighborhood to be preserved.

African Americans' opposition to the foundation and extension of historic districts has also been mentioned (see also: Silver 1991, pp. 79). Clearly, the historic districts represent a further urban forum in which ethnic differences and the problems of minority groups come to the foreground. Many minority groups have no or insufficient political representatives to voice their needs.

We must not overlook here the effects of the new understanding of history and the importance of aspects of heritage in relation to historic preservation programs in the United States. In this context, historic preservation can be seen to act as a means of bringing under control and cultivating the many global, cultural, and local influences and powers that are evident in a city. Nevertheless, the historic districts can be planned and set up in such a way as to ensure an interconnection between different regions. These regions can be represented by an urban district, a county, a city's SMSA or by an entire region.[162] In addition to all the structures

[161] This is particularly true in the case of Savannah and Philadelphia; public authorities in Charleston are attempting to address this problem. Boston has been relatively successful in its proposals for a linkage program (to attempt to find a balance between structurally weak and structurally strong areas of the city) (see: Levine 1987, p. 119). These projects should work above all to counteract the insidious gentrification process (Schneider-Sliwa 1995, p. 587).

[162] The most important models include: the principle of diffusion, the reurbanization process according to the New Urbanism movement, the creation of a permanent cityscape, the process of deurbanization (urban sprawl) and bridging processes.

that the South has in common with the North, it is important that the latter make provisions to prevent catastrophes.

It is today unquestionably true that historic preservation should act to boost a city‘s tourism industry. The varying and specific nature of the architecture and the place attract regional, national and international tourists. This is above all true in the case of all four historic districts examined here. At the same time, conditions in the neighborhoods are being improved along with the image they project to the outside world. Conferences and festivals offer tourists an even greater range of activities. Several examples were mentioned in the analytical part of this study. The conflicting aims and interests of the tourism industry were discussed in detail within the analysis of Charleston. We might wish that, in some American cities, historic preservation were not merely a means of attracting tourists. The variety of methods adopted in this study has enabled us to approach the question of historic preservation and related issues from many different perspectives. At the present time, historic preservation can be seen to be a cycle, ever up-to-date with current politics.

The reasons for introducing historic preservation programs vary from city to city. All cities agree on one thing: historic preservation should support and communicate ".. the meaning of place and community identity .." (Hodder 1999, p. 439). In the same way, structures should be preserved according to the ideals of the Urban Design School.[163]

It is also to be commended that many communities make the aims of historic preservation their own. Grass root ideals are respected and the official character of historic preservation is at the same time broadened and completed through the inclusion of elements of public welfare, in the interests of which property may be repossessed. The displacement of weaker social groups through gentrification is still one of the most difficult side effects of historic preservation, even though it has, to a large extent, been identified and quantified.[164] At this point, it is relevant to mention the tendency towards exclusivity that has developed in the areas of Beacon Hill and Society Hill. Conflicts of interest and the overall position of cities in relation to urban structures are of particular note in this context. Urban organizations work to combat the often one-sided access to resources (available support and funds) through their own specifically implemented programs.

The differences that exist between the various systems of historic preservation clearly show that only the methods devised by "individual urban geography" (*individuelle Stadtgeographie*) (Denecke 1989) can show which projects have a chance of succeeding. Simple statistical analyses can lead to vital considerations being omitted. It is only through an interdisciplinary examination, taking into account the multiple methodologies that exist, that an overall judgment may be reached that can be used as a basis for new structural work within the historic districts or for reaching other decisions involving the historic districts.

---

[163] The school's main advocates are mentioned only briefly here: AIA 1965, Lynch 1982, Alexander 1987, Garvin 1997.

[164] See further publications by Hodder (1999, pp. 439).

8.2 Extent of Generalization: U.S. American and German Findings Compared

The U.S. historic district concept is unique, taking into account cultural and planning concerns that cannot be compared with any existing concept in Germany or the rest of Europe. Historic preservation is not a research field with defined aesthetic norms or architectural methodology. The many facets of tangible structural and architectural history and the direct experience of living history[165] evident in the United States have long since been recognized by the old "culturally-concerned and culturally rich" world. Moreover, for this reason, the implementation of historic preservation in the United States is of particular interest to us as Europeans. Research into the people involved and their multiple motivations has led to a new understanding of alternative ways of introducing historic preservation programs. In this study, changes in the population and within society, as well as the touristic and economic value of historical structures, point to other questions that are equally relevant within a German context. International comparisons allow us to step back from national paradigms and images, widening our perspective and allowing us to identify other issues and work towards addressing them effectively.

If we consider in isolation elements such as definitions, concepts, long-term aims and architectural conversion, they appear superficially comparable. When analyzing these elements as a whole (meta-analysis), however, we must also take into account the highly complex structures and interdependence of the individual elements in question (Denecke 1989; "individual urban geography" - *individuelle Stadtgeographie*), all of which play a role within the field of historic preservation and urban planning. In this study, the main problem lies in the definition of the variables and the departure from traditional ways of thinking. The ideas and philosophies that underpin, for example, the administrative system, culture, existing buildings and internal urban structures are all determinants within this work. Germany on the one hand, and the whole of Europe on the other, are often compared to the U.S. in terms of scientific research, health and urban politics. Whereas Germany is often compared with other European countries and the U.S. is more often compared with England, the following is intended to bridge the gap between the two and permit a new comparison to be made.

8.3 Historic Districts: A Concept for Germany?

The descriptive terms applied to U.S. cities, namely private cities, individualistic cities und suburban cities, might equally be applied to the ever-growing cities in

[165] The term "living history" refers to a relatively new method in the United States: historic events are reconstructed in historic settings. The visitors' active participation plays an important role here (Murtagh 1997, p. 217).

Germany (and, indeed the rest of Europe).[166] These attributes can be transposed into a German context, even though they were originally identified in order to compare the United States and Canada (Table 8.1).

The results obtained in the analytical part of this study allow us to identify questions relevant to Germany. Although Germany and the United States are both developed capitalist societies in which many structural similarities may be identified, we might also go so far as to assume that an equally high degree of similarity exists in the field of historic preservation. The two countries have much in common, just as we can identify their differences according to their individual problems and potential. In the following section, the distinctive features that separate the U.S. and German urban system will be exposed and discussed.

Table 8.1. Urban attributes in the United States and Canada/Europe (Source: Ott 2000, adapted)

USA	**Canada/Europe (applicable to Germany)**
Private City	Public City
Individualistic City	Community City
Suburban Culture	City Culture City

[166] See lecture by Ott (1999), *Deutscher Geographentag Hamburg*, 1999.

Table 8.2. Distinctive features of U.S. and German cities

	USA	**Germany**
Urban population	77%[a]	87%[b]
Population density (general)	No lack of space, no dense construction	Lack of space, dense construction
Structural features of cities	No dense construction, depopulation	Dense population, increasing population
Overdevelopment of historical structures	Limited overconstruction[c] due to suburbanization	Marked overconstruction due to suburbanization
Location of jobs	Since the 1970s, increasingly to be found in the suburbs	Mostly in the city centers
Suburbanization	"Normal" way of life, "suburbian" lifestyle	Rare
Overall awareness	"City" lifestyle questioned	Preferred lifestyle
Relationship tenants vs. house owners	High number of owners (particularly in the historic districts), possibly leading to a more rapid gentrification process	Much higher number of tenants than owners, comparably greater tenant rights
Integration of historic preservation measures	Primarily by setting up historic districts, few individual buildings (often listed in the National Register)	Primarily individual buildings, few ensembles or other protected areas

[a] Source: Hofmeister (1999) and Fischer Weltalmanach 1999, p. 835.
[b] Source: Fischer Weltalmanach 1999, p. 163.
[c] With the exception of Prime Cities and Global Cities (see for this concept: Sassen 1991 and 1994).

In the 20th century, no specific authority worked to guide and coordinate state urban planning in the United States This contrasts starkly with the history of European urban planning. For this reason, urban planning in Germany to this day is profoundly influenced by the past, whereas the United States are free to deal with planning issues without this additional burden. The very phenomenon of "suburbia" (Lichtenberger 1986, p. 394) soon rendered plans for converting the existing urban structures obsolete.

Unlike American architects, European architects traditionally draw up very precise plans. This has made it much easier to set up archives and carry out architectural research into historically preserved structures in Europe and has furthered the movement to a large extent (Webster 1981, p. xxiii).

In 1995, for the first time, the German Committee for Historic Preservation (*Deutsches Komitee für Denkmalschutz*) (1999a, p. 15) published the following eleven criteria in order to define the general social and economic aims of historic preservation within Germany. According to these criteria, historic preservation consists of:

1. A modern approach to the politics of culture
2. The preservation of the cultural infrastructure
3. The preservation of different lifestyles
4. The achievement of locational quality
5. The support of the middle-class economy
6. The avoidance of costly and ill-advised investments
7. The support of the philosophy of thrift
8. The liberation of investment possibilities and the creation of jobs
9. A responsible use of public funds
10. Critical awareness
11. A reinforcement of social peace.

Several of these aims and theories can be identified in the context of historic preservation in the historic districts. A critical comparison shows us that, in Germany, the socio-cultural aspects of historic preservation assume pride of place (criteria 1 to 4 and 9 to 11). Only five of the eleven criteria address economic issues in the widest possible sense (criteria 5 to 8). It can be seen, above all, that different interests affect users and owners in the field of historic preservation in Germany.[167]
While the United States was undergoing suburbanization, the German population was also experiencing change that affected the function of cities as a result of increasing technical advances and greater mobility in the outlying suburbs of the cities (Kunze and Kurth 1998, p. 3).[168] The term "revitalization", which was for many years applied exclusively to American inner-city areas, is now widely used in connection with German inner-city areas. There too, much structural material remains, such as unused industrial structures and empty buildings from the industrial era. Converting these buildings into residential blocks is not always the only solution, as the creation of new jobs within the urban region is equally important.

Table 8.3. Features of users and owners with respect to German programs of historic preservation

Features of users	Features of owners
Particularly high locational quality compared to other buildings	Can rent at higher prices, better possibilities for sale of objects
Historical architecture is the hallmark of quality	Can rent at higher prices
Companies can develop a corporate identity	Users stay longer in one building because they identify with it
Modern interior decoration, providing conversion is not excluded or restricted	Profitable renting of property (good return on capital, maximum profits)

[167] The interests that are involved in the U.S.A. have already been identified, categorized and listed in table form in Chapter 3.

[168] Suburbs here refer to urban areas.

A judgment reached by the German constitutional court (*Bundesverfassungsgericht*) on 1st July, 1999 shows that some changes may be implemented within the system of historic preservation in the future. The court in Karlsruhe revealed a whole new dimension to the ideas of property and preservation responsibility: regulations concerning historic preservation should also take into account the interests of the property owner.[169] In Germany, historic preservation has traditionally been more highly regarded than in the United States and can be used to justify restrictions on property rights (Hofmeister 1982, p. 483). This new decision resulted from a case focusing on a 19th century villa that had been classified officially as a listed building since 1983. Due to the high cost of its maintenance and its lack of practical function, the court decided that the owners could not reasonably be expected to find a practical use for the building and decreed that it might be demolished.

Today in Germany, the preservation of historical buildings through private means, particularly in the case of real estate, is as endangered by these private individuals as it is dependent on them. Many private investors in Germany are put off by the complicated paperwork and endless formalities that must be completed before an historic object can be acquired. Fear of high costs, planning-related uncertainty and overly inflexible regulations regarding historic preservation are some of the main reasons behind this (Lorenz 1999, p. 12). Yet there is a current tendency in Germany to regard historically preserved real estate as a sign of quality in itself and such buildings develop their own locational criteria and qualities, causing prices to rise sharply, especially for buyers and tenants who decide to invest at a later stage. This tendency can only rarely be seen in the United States, the priority there being an immediate return on capital. The growing number of locations that are becoming independent of their surroundings is the result of new advances in communication technology, and this in turn can reignite interest in historical buildings and lend them new significance. This is particularly true in the case of old industrial complexes, which are often situated in remote locations. As far as the United States is concerned, we cannot make such generalizations, as not all Americans are interested in old buildings, especially those situated within inner-city areas. Even taking into account socio-historical attitudes, there are nevertheless some historical periods that are more or less ignored, as has been shown in the analytical part of this study. Private investors in Germany have different interests to state and local authorities and are interested in a dynamic approach to historic preservation, in other words, they want to use the old structures and reflect their original function, as this is the only way that these buildings can become profitable. Hamburg, for example, is a German state (*Bundesland*) where the local authorities have, in the past, set an example for others by their flexibility and open-handed planning (Lorenz 1999, p. 13). One of the most famous preserved buildings is the Chilehaus, where the facade, dating from 1924, has been preserved entirely, whereas the interior architecture is ultra-modern and dates from the 1990s.

---

[169] From the newspaper *Frankfurter Rundschau*, 2nd of July 1999.

There is evidence to show that Germany was also affected by a "historic preservation crisis" in the 1950s. During this period, old structures were often thoughtlessly demolished. As in the United States, some buildings were not completely demolished due to a lack of funds. Some historical buildings survived as a result of this, for which we can today, of course, be thankful.

In Germany, historic preservation is one of the state's major cultural concerns, making the existence of a partnership relationship between religious and private institutions all the more important. A series of planning regulations have been created in Europe (*Raumordnungsprogramme*) that have been classified in hierarchical order (Schneider-Sliwa 1995, p. 586). It is possible to generalize by saying that historic preservation in Germany is carried out "from the top down", whereas in American it is implemented "from the bottom up". This can be concluded from the American saying: "Less government is more government" (Staubach in: ILS 1999, p. 49). Historic preservation work is usually carried out in Germany with the public in mind and aims to provide a communal function. For this reason, historic preservation in Germany represents a unique and extremely sensitive area of urban planning, as it must simultaneously give a sense of identity and satisfy the current needs of the general urban population. In America, this is rarely the case. Historic preservation, as can be seen from the examples of Society Hill and Beacon Hill, is more a social tool and can clearly influence political decisions, whilst also reinforcing existing structures. On the other hand, historic preservation, as the examples of Charleston and Savannah have shown, can also work towards integration and support the existence of heterogeneous structures.

Quality of accommodation, especially in the cities, is also evaluated differently in Germany compared to the United States This is a factor that of necessity plays a role when considering the conversion of buildings. Quality standards and the demands of potential homeowners or tenants vary significantly between Germany and the United States. Home ownership and the social ideal of becoming a homeowner are much more common concepts in the United States than in Germany.

The qualitative process of selection that decides what is worthy of historic preservation and what should be converted or demolished is much more strictly regulated in German cities than in the United States. The highest authority that can assess whether a building or ensemble should be historically preserved or not is, in Germany, the state itself. As a result, urban structures are part of a more universal planning system. In the long term, this means that German cities are far more homogenous and this, in turn, means that the tendency towards segregation and exclusion is much less marked.

More recent publications in Germany, including the guidelines set down in the government's Program for the Future 2000 (*Zukunftsprogramm 2000*), which addresses issues of housing and urban construction, represent attempts to allow these two forms of construction to take place harmoniously. Within the German economic context, a housing benefit reform (*Wohngeldreform*) enables measures to be taken to modernize housing and help individuals to own their own apartment. The social focus of the German economic context is unheard-of in the United States and, from an ideological and social perspective, also unwanted. These integrated and universal economy-oriented provisions in Germany represent much more than a mere infrastructural policy; they are a part of a wider social policy (ID 1999, pp. 150). This German policy of simultaneously taking into account subject and object has clear political and social objectives and aims to integrate disadvantaged groups through investing in social issues and ensuring that structural improvements are provided for in urban programs. Further national programs such as the Urban Construction Promotion Program (*Städteaufbauförderung*) and the Social City Program (*Die Soziale Stadt*[170], in existence since July 1999), have been implemented with these aims in mind.

Compared to the United States, American-style cultural and social changes have only been introduced in German cities over the past few decades. Such changes refer specifically to socially selective and segregationalist tendencies. Expert opinion varies greatly as to whether segregationalist tendencies are also apparent in German cities or whether the increase in the number of ethnic groups in certain urban areas allows for a greater freedom of choice. According to how they choose to view the situation, the cities work either to counter or to promote this tendency. The fact remains, however, that there is a need for a policy of integration that does not destroy urban society. Planners can use areas that could potentially be classified as historic districts within Germany as a strategic means of ensuring the urban integration of various cultures and ethnic groups. A plan that might be seen to address this problem is the district organization plan (*Quartiers-management*) that has been introduced as part of the Social City program (*Die Soziale Stadt*).

In Germany, historic preservation is, on the whole, organized and implemented according to fixed criteria, which is not the case in the United States. German historic preservation authorities have created an inventory that lists individual objects. The German historic preservation movement has developed over a number of decades; this does not, however, mean that the aim here was to gain social acceptance for the movement, as it has always been regarded more as a kind of tradition. As a result, it is not relevant to attempt to define the various patterns of historic preservation within Germany in the same way that this study defined such patterns within the United States.

There are marked differences between German and U.S. historic preservation programs. These are summarized and compared in the following table.

---

[170] This program aims above all to promote local trade opportunities, active participation in decisions on new planning projects and private, economic involvement within the district (*Quartier*) (Nahr 1999, pp. 18).

Table 8.4. Historic preservation: Germany and the United States compared

Characteristics	United States	Germany
Extent of historic preservation	Historic districts/preservation of ensembles (predominantly)	Individual objects/preservation according to fixed criteria (predominantly)
Administrative features	Initiated by private institutions "from the bottom up"	Initiated by public institutions "from the top down"
Social motivation	Increasingly promoted, widely accepted since the 1970s and 1980s	Tradition (history, enlightenment and education as part of general social values)
Source of funding	Mostly privately funded, donations, trust funds, tax benefits available	Mostly publicly funded, indirect funding possible as a result of tax benefits
State power	Difficult for the state to intervene when private property is involved	Easy for the state to intervene to repossess property and agree on aims
Legislation	Many laws exist on a local level	Many laws exist on a national level
Historic preservation crisis	Took place in the 1950's, cities were revitalized, new buildings constructed in the suburbs, modern constructions became the new ideal, historic preservation was not yet widely accepted	Took place between 1950 and 1965, many buildings were demolished thoughtlessly and new construction took place. Little social acceptance in that period, money was lacking at times, owners unwillingly restored older structures
Approach towards historical features	Concentrated on return on capital, state was involved to some extent in cultural protection schemes	Cultural schemes, state's duty seen to be the continuing preservation of culture
Social acceptance	Growing, promoted by the state	Generally very high
Traditions	Only systematically implemented since 1935 through HABS/HAER	Measured plans traditionally available

Characteristics	United States	Germany
Regulations	Local authorities decide the extent historic preservation should be integrated in construction plans, precise laws exist above all in the historic districts in the form of the historic zoning ordinance	Historic preservation laws are incorporated into construction plans as a matter of course
Historic districts	An existing and active element of historic preservation in the United States according to law, in some cases non-binding design guidelines are available	Few and far-between, design guidelines are always set down by law
Historic preservation registers: national level	Available on a national level: the National Register of Historic Places	No national registers (exception: registers forming part of UNESCO programs)
Historic preservation registers: local level	Cities have their own local registers, though some leave much to be desired.[a] Several neighborhood organizations have voluntarily listed their historical features and made this information available to the city.	A generally well-organized register is kept by the individual states (*Bundesländer*)

[a] In many cases, these registers are based on so-called windshield surveys, in other words, assessments have been made from a moving vehicle (Iv-Bos-1).

Once a comparison has been made, the question arises as to whether the U.S. concept of historic districts can be applied to German cities in the future, even though the basic German system of historic preservation is significantly different from the American system. Hence, we must ask how they may be implemented within a German (or a European) context. The way in which historic buildings and areas are presented in the U.S. historic districts can be seen to be extremely progressive. Can American measures and structures be transposed into a European context? The organizational theories and structural forms used within the historic districts can undoubtedly be implemented in part in European planning strategy. Historic districts are today widely accepted within the United States and worldwide (Hodder 1999, p. 439). As the historic districts tradition dates back further in the United States than in Europe, some of its precepts can undoubtedly be adopted. Strong regional identification (for example, with a specific district or, in a German context, *Quartier*) or the German concept of district organization (*Quartiersmanagement*) can profit from these projects. Who is responsible for and whom such programs in Germany implicate can be seen in the following diagram.

The discussions surrounding urban and district organization (*Stadt- und Quartiersmanagement*) (Alisch 1998) can be structured according to the American example of citizen initiative in the historic districts. In his monograph, Rucht (1994) identifies structures that resemble those already existing in neighborhoods.

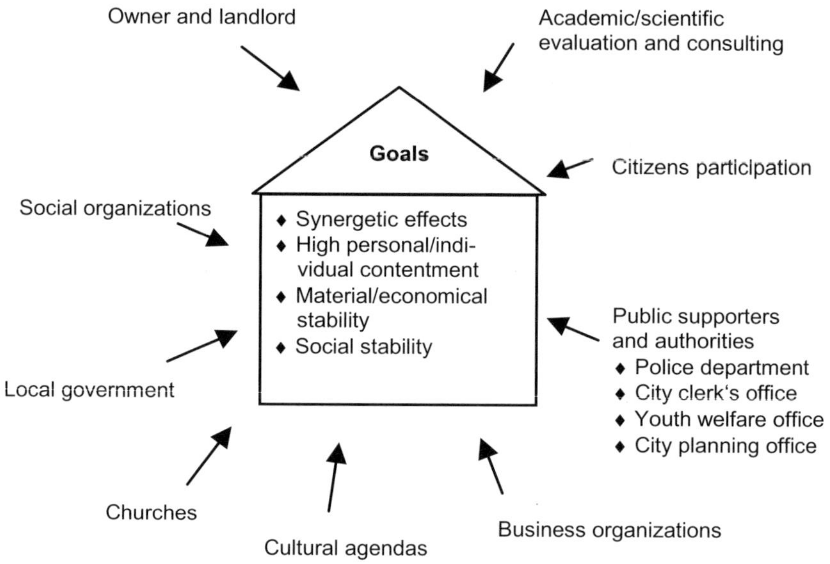

Networking among the representatives demands a high standard from the district management (*Quartiersmanagement*).

Figure 8.1. District (*Quartier*) Life

Table 8.5. Comparison of relevant progressive social movements (Source: Rucht 1994, p. 151)

	Civil liberalism	Worker movement	New social movement
Economic motivation	Absolutist despotism	Economic impoverishment, political control	Existential threat, cooperative activities, marketing of everyday life
Aims	Economic and political freedom	Social justice, political democracy	Preservation of essentials and quality of life, general participation
Social nucleus	Educated middle class	Skilled workers	New middle class
Internal organization	Cooperation, clubs	Association with specific aims	Group initiatives, networks
Strategy	Political revolution	Political revolution, nationalization of production	Radical reform, general implication of citizens
Model	Liberal state under rule of law	a) Democratic socialism b) Democratic capitalist welfare state	Unclear (active society, civilian society)

District organization (*Quartiersmanagement*) is implemented through the mobilization of local people and institutions within the districts. These self-defined forms of organization should enable inhabitants and workers to develop their identity and identify with cities on a local level. In this context, it is useful to refer to district-specific organizational systems, shown here in the following diagram.

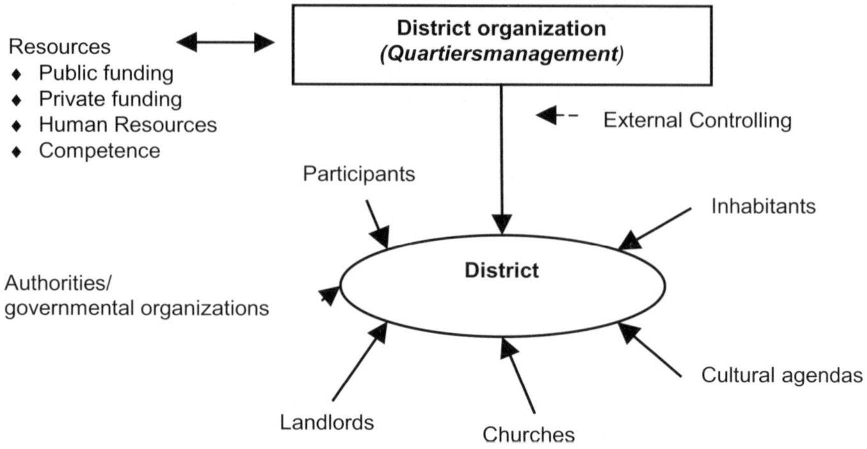

Figure 8.2. District organizational systems in Germany (schematic representation)

These long-term aims are, to some extent, comparable with existing structures within the U.S. neighborhood organizations. The German Social City Program *(Die soziale Stadt*) aims to improve both the cities' architectural and social structures.

It could also potentially be interesting if Germany were to redefine, for example, certain cities, areas or new regions[171], with slogans reflecting their identity. This is originally an American concept and has emerged as a result of the economic theories of urban marketing. In this way, the market value for the tourist industry is increased. Yet Europe has itself implemented a Cultural Capital Program (*das EU-Programm Kulturhauptstädte*). Some German cities have also been given unmistakable slogans to go with their names.[172]

Moreover, it is certainly feasible that a system of trails could be set up in Germany.[173] The model of historical stage-management in the case of "living history" can also be adapted to fit into a German context. What is more, serious thought should be given to the idea of implementing historic preservation programs in some of the larger regions. This would have the advantage of encouraging tourists to stay for longer periods of time. In this way, the tourist has a wider choice of activities, while tourist potential is increased because visitors would come from further away and stay in the area for longer periods. Furthermore, this would be advantageous to the concept of historic preservation itself, broadening its scope and increasing its impact.

[171] This refers to a wider region, such as an area of particular historical significance.

[172] Hamburg: the Gateway to the World (*Tor zur Welt*), Bonn: Federal City (*Bundesstadt*), Celle: the Half-Timbered City (*Fachwerkstadt*), Hannover: the Expo City (*Expostadt*), Christmas City Nürnberg (*Weihnachtsstadt*).

[173] The German term "historischer Lehrpfad" (historic trail) is here unnecessary, as it originally refers to a process of instruction (*belehren*). The concept of "living history" involves a process by which history is relived or experienced directly.

8.4 Conclusion

The various types of historic preservation and how they have been implemented within the historic districts in Philadelphia, Boston, Charleston and Savannah have been discussed in this study from both a practical and a theoretical perspective. At the same time, the many facets of the concept of historic preservation have been emphasized. The analysis within this work has shown that the concept of historic preservation is based on a wide variety of issues, including the individual cultural and historic reasons behind its foundation, as well as the different methodologies and motives involved. Since the concept of historic districts was first developed (in 1931, in Charleston), it has undergone significant changes, not only on a planning and conceptual level, but also as a result of the growing integration of ethnic minorities into the community, meaning that historic preservation is socially much more widely accepted within U.S. society.

On a theoretical and objective level, the following results emerge: a sharp rise in rents can be observed as a result of the massive gentrification process. Consequently, new state programs are being introduced currently to combat this tendency towards social exclusivity. This leads in turn to controversial discussions centering on the definition of social well-being. The state can decide which measures to take, but it may equally well choose to distance itself from state planning affairs and respect instead the maxims of private capitalism. Planning projects that aim towards integration are hindered by the segregationalist tendencies of American society, particularly within the urban historic districts. It is precisely in these districts that this problem arises as a result of a traditionalist attitude towards historical and cultural values, and the concern to ensure a certain quality of life. Although various microstructural processes may be observed in the historic districts, all show a clear tendency towards exclusivity. Only in the Savannah historic district is this tendency less marked.

On a methodological level, the following results can be seen: urban preservation should not be synonymous with stagnation; in other words, it is vital that it take into account modern social needs and issues. The vitality of a specific area must also have a regional impact. At the same time, it must be remembered that the buildings constructed today could be the monuments of tomorrow. Modern historic preservation movements work to combat ever-increasing social anonymity or construction plans that are too hastily carried out by investors in the neighborhoods ("local innovation", as Datel calls it: Datel 1985, p. 141).

The protection of the urban image is much more than the mere renovation of structural facades or the promotion of new investment opportunities. Equally important in the urban area or district (*Quartier*) is the promotion of programs intended to stimulate interest, channel such interest effectively and, at the same time, address key social issues. A city's image should not merely represent a ploy to attract tourists to the area, but should also reflect the urban population itself.

This can be seen to be problematic, in that newcomers and investors can, on occasions, reach irrational and unpredictable decisions that affect this urban image.

One of the most efficient ways of implementing programs of historic preservation is undoubtedly to make buildings and features of interest public property. It cannot be denied, however, that the historic district concept within the U.S. context of private capitalism, taking into account relevant local planning rights and measures, is a further example of how historic buildings may be preserved effectively. The U.S. neighborhood organizations can be seen to play a dominant and decisive role here.

It has been demonstrated that the historic districts today represent a multifacetted and useful urban planning tool. Historic districts are a means of protecting historical structures, whilst at the same time redefining their value and function. This study's interdisciplinary approach has enabled us to take into account a wide range of information from the various areas of specialist research related to the research field of historic preservation. This interdisciplinary focus is important, as it allows key information affecting urban historic preservation to be compared and combined (demographic information, segregationist tendencies, urban vitality, urban image from an external and internal perspective).

The comparison and transposition of these results to a German context has demonstrated that political, economic and socio-cultural factors must simultaneously be taken into account, meaning that processes may only be compared to a certain extent. An equivalent comparison can only be made in a German context according to a specific set of criteria, as has been shown, for example, by the way in which Germany has implemented the concept of "living history".

Zusammenfassung: Einleitung

Die etwa 500jährige Zeitspanne, die seit der Entdeckung des amerikanischen Kontinents verging, ist, gemessen an der europäischen Besiedelungsgeschichte zwar relativ kurz, dennoch verfügt die US-amerikanische Nation über eine Vielzahl von historischen Gebäuden aus dem 16. bis 19. Jahrhundert. Gerade in den Städten der dreizehn Gründerstaaten und in den Städten der Südostküste der USA gibt es heute eine bemerkenswert hohe Zahl und Dichte an noch erhaltenen und erhaltenswerten historischen Bauwerken. Deren Restaurierung, Rekonstruktion, Umnutzung sowie deren nachhaltige Pflege müssen als ein wichtiger Beitrag zur Pflege des baulichen und kulturellen Erbes der USA gesehen werden. In diesem Buch werden die Form, das Ausmaß und die Motivation US-amerikanischer Denkmalpflege identifiziert und analysiert.

Seit den 1960er Jahren wird in den USA der thematischen Verbindung von Denkmalpflege und Stadtplanung ein zunehmendes Maß an gesellschaftlicher und wissenschaftlicher Aufmerksamkeit gewidmet. Geplante städtische Denkmalpflege ist heute mehr als nur eine „Sehnsucht nach Geschichte", sie ist ein aktiv eingesetztes Planungsmittel in den Städten der USA. Setzt man sich mit amerikanischer Denkmalpflege im städtischen Raum auseinander, stößt man schnell auf das Konzept der historic districts, dem eine besondere raumprägende Relevanz zukommt. Um diesen stadtplanerischen, ökonomischen und gesellschaftlich-prozessualen Bedeutungen nachzugehen, sind empirische Untersuchungen in vier städtische historic districts zum Forschungsgegenstand dieser Arbeit vorgenommen worden. Als regionale Beobachtungseinheiten sind sie in den Städten Philadelphia, Boston, Charleston und Savannah gewählt worden, die historisch bedeutende Hafenstädte an der Ostküste der USA waren und die im ehemaligem Kolonialgebiet von England liegen.

Alle untersuchten historic districts befinden sich innerhalb alter Stadtkerne und sind Bestandteile der heutigen Central Business Districts (CBD). Die ausgewählten Mittel- und Großstädte sind von überregionaler Bedeutung, haben aber keine Sonderfunktion innerhalb der USA. Der Sezessionskrieg (1861-1865) bringt bei näherer Betrachtung jedoch einen interessanten Kontrast in die Städte der Nord- (Philadelphia und Boston) und Südstaaten (Charleston und Savannah), der sich im Rahmen der Fragestellung deutlich akzentuieren lässt.

Unter Verwendung interdisziplinärer Methoden soll aufgezeigt werden, daß der Denkmalschutz innerhalb der historic districts ein neues Licht auf die Dynamik des städtischen Raumes der USA wirft. Dabei wird die planerische, ökonomische und gesellschaftliche Komplexität des Konzeptes herausgearbeitet.

Welche Ziele, Motivationen und Planungsaspekte innerhalb der historic districts der USA eine Rolle spielen, bildet die Forschungsgrundlage dieses Buchs. Innerhalb der amerikanischen Städte spiegelt sich die Präsenz unterschiedlicher Ethnien, Generationen und sozialer Schichten wider, die eine hohe Konzentration mehrerer Interessen- und Raumkonflikte aufweisen. Diese gesellschaftlichen

Wohnumfeldbedürfnisse umfassen einen großen Bereich an Identitäten und planerischen Parametern im städtischem Raum, die auch in den historic districts ihren Ausdruck finden. An konkreten Motivationen und Interessen und nicht an einer fiktiven Stadtgesellschaft orientiert sich darum der empirische Teil weil die Prägung und Nutzung von Räumen nicht nur physischen, sondern ebenso sozialen Determinanten unterliegt. Es geht darum nicht nur um die rückwärtsgewandte Betrachtung der Umsetzung von Denkmalpflege, sondern vielmehr um die praxisorientierte Planung der historic districts und die ökonomische Inwertsetzung der Zukunft (Stadtgestalt) der Städte. Die zentrale Fragestellung der Studie lautet dabei:

Wie lassen sich Kultur und Geschichte durch historic districts gleichzeitig angemessen zeitlos und zeitgemäß in das gegenwärtige Stadtgefüge und die gesellschaftliche und ökonomische Stadtvitalität einbinden?

Gegensätzliche Thesen, Theorien und Umsetzungsideen stehen dabei im Vordergrund: Historic districts können einerseits als planerisches Instrument eingesetzt werden, um homogene Bevölkerungsgruppierungen innerhalb eines Bezirks zu vermeiden. Historic districts können jedoch andererseits auch als lenkendes stadtpolitisches Mittel eingesetzt werden und damit gesellschaftliche und soziale sowie ökonomische Integrationsarbeit leisten.

Grundsätzlich lassen sich die vorliegenden Publikationen im Bereich der Denkmalpflege chronologisch und thematisch in sechs Bereiche einteilen:

1. Vom Beginn des 20. Jahrhunderts und bis zum Beginn des Ersten Weltkriegs
2. Um 1900 kam im Rahmen des City Beautiful Movement
3. Seit 1950 im Rahmen der massiven Suburbanisierung
4. Seit 1960 Untersuchungen zu Stadtplanungsvorhaben und -empfehlungen einiger Städte
5. Ende der 1970er ökonomische Aspekte der Denkmalpflege und deren Potentiale wurden analysiert
6. Seit einigen Jahren: touristische Sponsoring, Städte im globalen Netz der Urbanisierung, Steuerdiskussionen, Denkmalpflege und Erziehung sowie der Aspekt des heritage

Diese sechs Bereiche zeigen, dass sich das allgemeine Interesse an Architektur und Geschichte in der öffentlichen Meinung in den letzten Jahren stark geändert hat. Eine Bewusstseinsweckung ging unter anderem auch auf die Civil Rights-Gesetzgebung von 1965 zurück.

Zusammenfassung: Methodik und Terminologie

Für die Analyse der vier historic districts mit ihren Organisationsstrukturen, ihren Interessenfeldern und ihren Potentialen müssen zunächst die theoretischen und administrativen Grundlagen der Denkmalpflege in den USA erforscht werden. Dazu müssen entsprechende Mikro-, Meso- und Makroebenen des zu beforschenden Feldes ausdifferenziert werden, die dann die räumlichen Beobachtungseinheiten bilden.

Je nach Untersuchungsmaßstab können dann die Interessenkonflikte im Rahmen der denkmalpflegerischen Maßnahmen festgestellt werden (Raumnutzung, ökonomische Aspekte, Stadtbild, Stadtgestalt, Außenbild- und Innenbild).

Der Methodenpluralismus setzt sich aus Primärerhebungen wie Experteninterviews, Gebietskartierungen und -beobachtungen sowie aus Sekundäranalysen zusammen. Dieser Mix ist erforderlich, um ein Gesamtbild der historic districts herauszuarbeiten.

Statistisches Material und Planungsmaterial bilden die Sekundärquellen. Insbesondere sind hier die Darstellungen der räumlich-sozialen Verteilung und der Festschreibung lokaler denkmalpflegerischer Vorschriften zu nennen (historic district ordinance, historic zoning ordinance). Weiterhin wurden Veröffentlichungen des US-Census, der nationalen, einzelstaatlichen und lokalen US-Behörden, der einzelnen denkmalpflegerisch tätigen Institutionen (beispielsweise NTHP - National Trust for Historic Preservation oder AIA - American Institute for Architects) für die Analysen herangezogen. Organisationstrukturen der neighborhood-Organisationen oder anderer Institutionen liefern weitere Quellen, deren Wortlaut zum Teil ausgewertet und graphisch aufgearbeitet wurde. Schließlich runden Baunutzungskarten, Mitgliederzeitungen, Satzungen und lokale Presse diesen Quellenfundus ab. Die Sekundäranalysen erschließen eher Strukturdaten, während die Primärquellen auch zur Klärung der Handlungsdimension innerhalb der historic districts herangezogen werden.

Die Paarung des theoriegeleiteten und des analytischen Ansatzes im Rahmen dieser Untersuchung repräsentiert ein induktives Verfahren, der Teilaspekte in einer differenzierten Auswertung genau untersucht, um schließlich Rückschlüsse auf das gesamte Themengebiet der städtischen historic districts vorzunehmen. Die statistische Gesamtaufnahme einer neighborhood ist aus diesem Grund exemplarisch nur für den historic district Society Hill in Philadelphia vorgenommen worden.

Zusammenfassung: Denkmalpflege und Stadtplanung

Eine Arbeit über Denkmalschutz erfordert es, einige grundsätzliche Aspekte von Geschichte und Historizität zu erläutern. Die forschungsleitenden Fragen bei der Interpretation von Geschichte und historischen Quellen sind die gleichen wie bei (denkmalgeschützten) Gebäuden:

1. Woran wird erinnert und welchem Auswahlprozess folgt dabei die Dokumentation?
2. Warum bleiben uns bestimmte Dinge erhalten und welche Zwischenstücke fehlen uns möglicherweise?
3. Welche Form der Reflexion von Realität soll uns vermittelt werden?

Es ist heutzutage nicht zu übersehen, dass Begriffe wie „heritage“ und „historisch“ in den USA sehr häufig verwendet werden. Vielleicht ist es eine gewisse Zukunftsscheu, welche die Gesellschaft wieder zurückblicken lässt. In der Unüberschaubarkeit der Moderne kann die Überschaubarkeit der Historie, die in der Begrenztheit der Daten begründet ist, das bestehende Bedürfnis nach Identität leichter festmachen. Einige Kritiker behaupten sogar, dass gerade in Deutschland und Europa eine Geschichtsfixierung herrscht, die dem allgemeinen Trend des Zukunftspessimismus entgegenwirken will. Indem Denkmalschutz in den historic districts zu einem originären Problem der Stadtplanung wird, lässt sich daran auch der Glaube an Kultur festmachen und sogar manipulieren.

Inwiefern Gebäuden oder auch den Leitbauten eine kulturelle Bedeutsamkeit zukommt, obliegt Einzelfallinterpretationen. Ob einige Leitbauten auch zu „Leidbauten“ werden können, weil sie an eine bestimmte Epoche unangenehm erinnern, soll dabei untersucht werden. Die Historizität eines Gebäudes kann jedoch auch durch eine verfremdete Erhaltung in seinem Umfeld die historische Ureigenschaft und auch Aussagekraft verlieren. Diese Diskrepanz zwischen der Esoterik (dem wissenschaftlichen denkmalpflegerischen Anspruch) und der Exoterik (der gleichzeitigen „Lizenz zur Verständlichkeit") ist von den entsprechenden Akteuren des Denkmalschutzes zu erkennen und durch entsprechende Maßnahmen aufzufangen. Funktionalistische Überlegungen konkretisieren dabei „Kultur“ und verbinden das Gebäude dabei zu einem Baudenkmal. Um alle ethnischen Gruppen in diese Betrachtung einzubeziehen, muss dann auch noch die Bedeutungskategorie für die entsprechende Gruppe identifiziert werden.

Ähnlich wie der deutsche Begriff „umweltfreundlich“ unterliegen die Begriffe historic preservation, heritage und landmark keinen offiziellen Definitionen. Dementsprechend muss für die objektive Beurteilung des Denkmalschutzes das Bezugssystem hergestellt werden.

Wie wird unter diesen Voraussetzungen Denkmalpflege im Privatkapitalismus vorangetrieben und propagiert? Denkmalpflege sollte nach Fitch so gestaltet werden, dass sie langfristig ökonomische Vorteile und eine Revitalisierung des Bezirkes mit sich bringt. Hier ist insbesondere an den Tourismus, an allgemeine Aufwertungsmaßnahmen und an die neue Branchen (Internet, High-tech) gedacht worden. Der ökonomische Profit hängt also vom Grad der Förderung und dem Betreiben der Denkmalpflege ab.

Der Begriff „heritage“ wird nicht als einziges oder alleiniges raumprägendes Element verstanden. Die Begriffe „Denkmalpflege“ und „Denkmalschutz“ hingegen beziehen sich stets auf baulich existente und physisch tastbare Elemente wie zum Beispiel Gebäude, während heritage eine wichtige Erklärungsdimension der Unterschutzstellung liefert.

Die unterschiedlichen Motivationen, Denkmalpflege zu betreiben, wurden verschiedene Themenkomplexe der Stadtplanung herausgearbeitet, isoliert und diskutiert. Dabei spielt vor allem die Verknüpfung der beiden Themenkomplexe Stadtplanung und Denkmalpflege die zentrale Rolle.

Zwischenergebnis: Die Historic Districts im städtischen System der USA

Politisch-administrative Beurteilungsebene

In diesem Kapitel wurde aufgezeigt, daß der Grad der Denkmalpflege in den Städten der USA sehr stark von der lokalen Regierung abhängt. Denkmalpflege gestaltet sich als eine sehr lokale und private Angelegenheit, bei der die Regierung in eingeschränktem Maße geringe Initiative, Unterstützung und Förderung leistet. Es sind vor allem die lokalen Akteure und Mikroprozesse in den Stadtbezirken und neighborhoods, welche die Denkmalpflege maßgeblich anschieben.

Die Bemühungen des Staates im Bereich der Denkmalpflege garantierten bis 1974 keinen wirklichen Schutz. Eigentümer historischer oder signifikanter Häuser waren lediglich dazu aufgefordert, ihre Häuser eigenverantwortlich zu erhalten. Erst auf lokaler Ebene wird mit größerem Einfluß über Veränderungen und Abriß von Denkmälern und Gebäuden diskutiert. Innerhalb der historic districts ist dieser Schutz noch deutlicher gewährleistet. Sogar Neubauten werden in den historic districts an die bestehende historische Substanz farblich oder durch Material- oder Formwahl visuell angepaßt.

Die existierenden Denkmäler in den USA statistisch wären sicherlich nicht so systematisch inventarisiert worden, wenn nicht die Regierungen durch Gesetze und staatliche Organe eingegriffen und die entsprechenden Ämter eingerichtet hätten. Es existieren zu viele Interessenkonflikte, die vor allem im Bereich des Tourismus, der Wohnbevölkerung (neighborhood) und der Investoren zu finden sind, so daß eine lenkende Funktion des Staates sehr hilfreich ist. Deutlich wurde jedoch, daß die lenkenden Maßnahmen sehr stark profitorientiert sind und (wie zum Beispiel die Listungen und Inventarisierungen des historischen Bestands) an die einzelstaatlichen und lokalen Behörden abgegeben werden. Staatliche und politische Kontrollen verhindern heute einige voreilige Abrisse oder bauliche Veränderungen an der historischen Substanz, bei der eine Rendite nicht garantiert ist.

Ferner ist es erforderlich, in den historic districts lokale Flächennutzungspläne (historic zoning ordinances) mit den entsprechenden Aktualisierungen (zoning ordinance adjustments) zu entwickeln, um neuen Entwicklungen und der Denkmalpflege mit ihren Interessen gerecht zu werden. Jedoch auch bei der Erstellung dieser Gesetze und Richtlinien wird partizipierendes Planen durch die Bevölkerung im Sinne der grass-root-Demokratie gefördert.

Ökonomische Beurteilungsebene

Es ist umstritten, ob die Erfolge der Denkmalpflege objektiv meßbar sind und eine Isolierung ökonomischer Auswirkungen möglich ist. Wissenschaftlich sind ihre Auswirkungen nur schwer beweisbar, weil in den Kernstädten der USA viele verschiedene Programme mit unterschiedlichen Zielvorstellungen im Bereich der Denkmalpflege greifen. Erfolgskontrollen sind darum besonders kompliziert, was der gesellschaftlichen Akzeptanz und der Bedeutung der Denkmalpflege in den USA nicht unbedingt zuträglich ist. Traditionalisten argumentieren mit Kultur und Geschichte, Entwickler sehen die ökonomischen Erfolge im Vordergrund. Oft kollidieren diese Interessen, was sich insbesondere in öffentlichen und privaten Motivationen widerspiegelt. Diese These von Cullingworth trifft jedoch nur bedingt auf die Dynamik in den historic districts zu, weil dort die öffentlichen und privaten Interessen oft deckungsgleich sind, während Privatpersonen versuchen, sich den zum Teil sehr scharfen lokalen Gesetzen zu entziehen oder neighborhood-Organisationen die erkämpften quality-of-life-issues gefährdet sehen.

Die US-Regierung hat im Jahr 1976 einen wesentlichen Motivationsfaktor in der Bevölkerung geschaffen, um die Denkmalpflege zu fördern. Der Tax Reform Act von 1976 (public law 94-455) etablierte wichtige Steuervergünstigungen für die Erhaltung und Rehabilitation von gewerblichen und einkommensteigernden Strukturen, wenn sie durch das U.S. Department of the Interior als historisch anerkannt sind. Es werden jedoch auch Strafen verhängt (zum Beispiel durch die Erhöhung der Steuersätze), wenn historische Strukturen (wissentlich) zerstört werden. Die Steuerabschreibungen veränderten die gesellschaftliche Akzeptanz der Denkmalpflege in den USA erheblich, weil auch private Projektentwickler ein profitableres Geschäft aus neuen, den Denkmalschutz integrierenden Plänen erzielen konnten. Vorher war die Gesetzgebung so kontraproduktiv, daß sogar die Zerstörung von historischen Gebäuden absetzbar war. Unter anderem bezog man sich in solchen Fällen auf die Gesetzgebungen des Urban Renewal Program.

Steuervergünstigungen im Rahmen des National Neighborhood Policy Act (Title II of public law 94-24) lassen weitere passive Schutzmaßnahmen. Dieser neue Paragraph schreibt fest, daß bestehende städtische neighborhoods eine nationale Ressource sind, die es zu erhalten und zu revitalisieren gilt. Diese Festschreibung hat vor allem Ende der 1970er Jahre eine weitere Motivationswelle im Bereich der historic districts. Ferner versuchen seitdem verschiedene Organisationen, einzelne heruntergekommene neighborhoods in den Städten mit Hilfe dieser Gelder zu revitalisieren. Um Fördergelder bemühen sich hier vor allem strukturstarke und gentrifizierte neighborhoods und unterschiedliche öffentliche Organisationen, die sich besonders die Förderung und Stützung der Minoritäten in den Städten auf die Fahne geschrieben haben. Beide Ansätze werden im empirischen Teil weiter thematisiert und exemplarisiert.

Ein weiterer Meilenstein im Bereich der steuerlichen Absetzbarkeit von denkmalpflegerischen Investitionen und damit der indirekten Förderung von Denkmalpflege war durch den vom US-Kongreß verabschiedeten Economic Recovery Tax Act im Jahre 1981 erreicht worden. Dieser Paragraph vergibt eine 25%-ige Steuerermäßigung für die Rehabilitation von historischen Gebäuden. Dieses Gesetz hat besonders in urbanen Gebieten die Denkmalpflege maßgeblich vorangetrieben. Der während der Amtsperiode von Präsident Reagan 1986 verabschiedete Tax Reform Act hat die Vergünstigungen stark zurückgestutzt, aber die Investoren nicht völlig abgeschreckt. Die Formulierungen und genauen Methoden variieren jedoch von Staat zu Staat erheblich. Fonds und Spenden von verschiedenen privaten und geschäftlichen sowie öffentlichen Stellen, die ebenfalls mit steuerlichen Vergünstigungen vergütet werden, sind eine weitere übliche Form, Denkmalpflege zu unterstützen.

Cullingworth kritisiert in diesem Zusammenhang sogar die Inwertsetzung der US-Parks, die seiner Ansicht nach in Gefahr geraten, zu Themenparks oder Zoos zu mutieren. Viele lokale Banken, die von den Mikrokonzepten der Revitalisierung im Rahmen der Denkmalpflege überzeugt sind, stellen Kredite zur Umsetzung von Einzelprojekten zur Verfügung.

Auch die in den Städten bis heute noch existierenden Redevelopment Authorities (RDA), die in den Zeiten des Urban Renewal (1950er und 1960er Jahre) gegründet wurden, haben in einigen Städten die Möglichkeit, mit lokalen Geldern Denkmalpflege im kleinen Rahmen zu betreiben.

Obwohl viele Programme, Möglichkeiten und Hilfsprojekte im Bereich der Denkmalpflege existieren, liegt die finanzielle Hauptlast nach wie vor bei den Investoren. Nur finanziell bessergestellte Investoren können Steuerabschreibungen gewinnbringend nutzen und gegebenenfalls auch geringe Verluste in Kauf nehmen. Noch in den 1970er Jahren wurde die Denkmalpflege in den USA als ein Alleingeschäft für einkommensstarke Personen gesehen.

Bereits 1977 wurde festgestellt, daß außer den rechtlichen Grundsatzfragen über Eigentumsrechte das Hauptproblem in der Beschaffung der finanziellen Mittel für die Instandsetzung liegt. Diese Wahrnehmung hat sich bis heute nicht wesentlich gewandelt. Da sich aber das Bewußtsein zugunsten der Denkmalpflege deutlich gewandelt hat, sind auch mehr Fördergelder und deren Beschaffung durch und mit Hilfe der vielen Akteure vereinfacht worden. Wie diese ökonomischen Komponenten gewinnbringend regional in den einzelnen historic districts eingebunden werden, wird in den folgenden Kapiteln diskutiert.

Sozio-kulturelle Beurteilungsebene

Die Beschreibung der Akteure im Feld der Denkmalpflege hat offengelegt, daß differenzierte gesellschaftliche Motive das Themenfeld der städtischen Denkmalpflege kennzeichnen. Insbesondere die neighborhood-Organisationen sind als einflußreiche Akteure für die historic districts zu nennen.

Zusammenfassend läßt sich zunächst festhalten, daß die Denkmalpflegebewegung in den USA im Rahmen der grass-root-Demokratie zunächst sehr unorganisiert wuchs. Die Bewegung ging zunächst von Einzelpersonen aus, die hauptsächlich ausgewählte einzelne Gebäude schützen wollten. Viele dieser Versuche schlugen jedoch auch fehl, weil die Finanzierungsmöglichkeiten fehlten. Aus diesem Grund ist zum Beispiel das John Hancock House in Boston im Jahr 1863 abgerissen worden. Die Independence Hall von Philadelphia entging diesem Schicksaal nur knapp, als das Gebäudeensemble 1816 für den damaligen Spottpreis von nur 70.000 US$ verkauft wurde. Die Nationalparks dagegen waren schon früh im öffentlichen Bewußtsein. Architektonisch-denkmalpflegerische Werte bezüglich der städtebaulichen Denkmalpflege kamen erst gegen Ende des 19. Jahrhunderts

dazu. Vorher hatte man sich eher auf die US-amerikanische Geschichte beschränkt und die für diese repräsentativen Bauwerke geschützt.
Im Jahr 1978 wurde aus diesem Grund das National Council for Preservation Education gegründet. Dieses Gremium vergibt unter anderem Gelder und Förderungen im Bereich der Verbreitung von Denkmalpflege, insbesondere gemäß ihren Zielen, ihren Idealen, ihrer historischen Entwicklung und Förderung. Damit soll die gesellschaftliche Kenntnis und Akzeptanz gefördert werden. Im Bereich der Schulbildung sind Arbeitsmaterialien (Arbeitsblätter, Bücher) erstellt worden, die nach Bedarf im Curriculum der einzelnen Staaten der USA eingesetzt werden können.

Denkmalpflege bietet ein neues Feld auch zur Erzeugung permanenter Kontroversen. Die Ortsgebundenheit und selektive Ortswahl einiger US-Amerikaner läßt die Denkmalpflege vorantreiben. Unterschiedliche Bewohner haben verschiedene Erwartungen an die neighborhood, in der sie leben. Einige interessiert nur die Erfüllung und Gewährleistung der quality-of-life-issues, andere wollen sich auch kulturell und damit denkmalpflegerisch die neighborhood eingebunden wissen.

Es sei noch darauf hingewiesen, daß der soziale Wert von historic districts schwer mit monetärem Wert gleichzusetzen ist. Dieser diffizilen Frage soll in dieser Arbeit auch nicht detailliert, sondern nur exemplarisch nachgegangen werden. Der erste Blick auf ein Bauwerk ist meistens emotional, weshalb die Denkmalpflege einen starken kulturellen und emotionalen Wert bedeutet, der einen bedeutenden Teil der heritage ausmacht. In diesem Zusammenhang unterscheidet drei unterschiedliche Werte: den politischen, den spirituellen und den symbolischen Wert.

Zusammenfassung: Philadelphia: Der Urban Renewal-Typ

Society Hill, in seiner Gesamtheit voll geplant, wird sehr häufig in der Literatur erwähnt. Wie vorher nachgewiesen wurde, muss Society Hill bezüglich der denkmalpflegerischen Leistung als Sonderfall angeführt werden. Die Umnutzungen des Mischviertels mit Lagerhäusern, Scheunen, alten Geschäften und die Eingliederung der alten historischen Substanz in moderne Wohnannehmlichkeiten ist heute ein herausragendes Beispiel denkmalpflegerischer Leistung in den USA, aber dennoch ursprünglich nicht als Denkmalpflege-Projekt gedacht gewesen. Das „Renommierprojekt" Society Hill, wurde von Denkmalschützern in den USA damals abgelehnt. Diese Meinung wird heute nicht mehr ganz aufrechterhalten, weil die Kritiker von damals einsehen mussten, dass sonst womöglich die gesamte neighborhood dem Abriss zum Opfer gefallen wäre. Ferner ist die Initiative der neighborhood-Organisation heute sehr überzeugend.

Naherholung, Grünflächen, Parkplätze und die Erreichbarkeit von Geschäften des kurz-, mittel-, und langfristigen Bedarfs sind in Society Hill als gelungen einzustufen. Nicht zuletzt kann dieser Erfolg dem Joint-venture öffentlicher und privater Institutionen, vor allem auch der neighborhood-Organisation zugeschrieben werden. Die Delaware Valley Regional Planning Commission schätzt für Center City eine Bevölkerungszunahme von 14,2% innerhalb der nächsten Jahre. Für alle anderen Bezirke ist eine Bevölkerungsabnahme prognostiziert. Diese Zahlen verdeutlichen die andauernde und zunehmende Attraktivität von Center City. Eine Ausweitung der denkmalpflegerischen Protektionsmaßnahmen und die weitere Zunahme der Wohnbevölkerung in Center City ist auch in Zukunft zu unterstützen. Das „suburbane Motiv" von Society Hill ist nicht zu übersehen.

Dennoch muß in diesem Zusammenhang auch auf die Kehrseiten der Erfolge des historic district hingewiesen werden:

Die (Re-) Kreation des schützenswerten historisch-kolonialen Philadelphia ist vonstatten gegangen, ohne die historisch authentischen Vorgänge des 19. und 20. Jahrhunderts zur Kenntnis zu nehmen. Alle nicht im kolonialen oder nachkolonialen Stil erbauten Gebäude wurden abgerissen und so regelrecht Straßenzeilen im Federal Style errichtet. Weiterhin wurden zur Verbesserung der Infrastruktur, das heißt zum Beispiel zum Bau von Straßen, historische Kolonialstilhäuser abgerissen. Insbesondere die Häuser auf der Flußseite des Hill fielen diesem Projekt zum Opfer. Wegen des Baus der Autobahn (I-95) am Hafen mußten mehrere Hafenanlagen komplett abgerissen werden. Die bis zu 16-spurige Autobahn verhindert bis heute die uneingeschränkte Erreichbarkeit des Delaware River. Aus diesem Grund wurde 1998 eine Fußgängerbrücke in der Verlängerung der South Street eingeweiht.

Ein weiteres Potential für Society Hill sind die im Laufe der Jahre zugeschütteten Flussbetten, die wieder freigelegt werden könnten. Zudem sind durch die RDA neue Gebäude errichtet worden, die lediglich von der Farbgebung, der Außenfassade und der Höhe mit den anderen Häusern dem Gebiet kongruent sind.

Nur die später errichteten Kirchen und Synagogen wurden erhalten und prägen heute die Silhouette von Society Hill. bezeichnet diesen Vorgang als Stilschrumpfung. Es existieren offensichtliche Unterschiede zwischen Erhaltung und Instandhaltung, die an dieser Stelle „graue“ Denkmalpflege genannt werden soll(t)en.

Es wurden typische und untypische Muster im Rahmen der Stadtrevitalisierung aufgezeigt. Philadelphia fungiert an dieser Stelle als Urban Renewal-Typ der städtischen Denkmalpflege in den USA. Es ist eine wichtige und richtige Entscheidung in den 1950er Jahren gefällt worden, die neighborhood in dieser Form zu sanieren. Nachdem die RDA die Gebäude im Rahmen des revolving fund wieder veräußert hatte und sich die neighborhood-Strukturen stabilisiert hatten, haben die PHC und auch die neighborhood-Organisation die Qualitätssicherung der neighborhood übernommen. Heute übernehmen oftmals Einzelpersonen die administrative Mikroarbeiten, wie zum Beispiel der Überprüfung, ob die Denkmalpflege-Vorschriften durch die Eigentümer eingehalten werden.

Zusammenfassung: Boston: Der Neighborhood-Typ

Seit 1955 besteht in Beacon Hill ein Schutz der denkmalpflegerischen Substanz im Rahmen der Gesetzgebung für einen historic district. Es ist deutlich sichtbar und außergewöhnlich, wie aktiv die neighborhood-Organisation den Bezirk erhält. RUTZ weist nach, daß partizipierendes Planen in den USA sehr effektiv und keineswegs marginal ist. Als besonders erfolgreiches Beispiel hierfür kann die Genese des historic district in Beacon Hill gelten. Jedoch auch nach Abschluß der Ernennung ist es heute eine große Herausforderung, trotz der historischen Bausubstanz die neighborhood für die Bewohner vital zu erhalten, aber auch Touristen zugänglich zu machen.

Die Konkurrenz zwischen Back Bay und Beacon Hill und neuerdings auch South End in Boston ist nicht zu übersehen. Häufig wird öffentlich verglichen und die historisch-architektonische Substanz verbessert. Mehrere stadtweit ausgeschriebene Denkmalpflege-Wettbewerbe werden in verbissener Konkurrenz ausgetragen. Ferner geht es auch um den Wettbewerb um öffentliche Fördermittel. Beacon Hill und Back Bay sind zwei neighborhoods, die für Zuzügler niemals ganz „unmodern“ gewesen oder wirtschaftlich verfallen sind. Dies macht nach bisherigen Erkenntnissen die Ausgangslage für die Errichtung eines National Historic District wesentlich einfacher. Zu diesem Schluß gelangt man nicht nur durch beobachtende Analysen der neighborhood, die Strukturdaten von Beacon Hill belegen die gleiche These. Ähnliche gentrifizierte Strukturdaten wurden bereits für Society Hill nachgewiesen.

Die gesamte Phase der Revitalisierung durch die Urban Renewal Projects, wie sie in Society Hill vollzogen werden mußten, fiel in Beacon Hill weg. Demnach ist auch keine neighborhood (in ihrem Gefüge) völlig aus dem Stadtgefüge herausgelöst worden oder mußte neu aufgebaut werden. In Society Hill war die neighborhood-Struktur auch durch die Suburbanisierungsphase und die Zeit während der Depression vollends zerstört worden, so daß eine externe Rehabilitierung notwendig gewesen ist. Diese bestehenden neighborhood-Strukturen und auch die massive Unterstützung durch die lokale Stadtbehörde trugen maßgeblich zu dem Bild bei, in dem sich heute Beacon Hill präsentiert.

Zusammenfassung: Charleston: Der Planungstyp

Von der Planung her gesehen, stellt sich Charleston als sehr gemischter Typ dar. Trotz vergleichsweise starker Einlenkung durch die lokalen Behörden, die seit 1931 Urban Design und Stadtplanung als Mittel nutzen, Denkmalpflege zu betreiben, gibt es in Charleston viele denkmalpflegerische Lücken. Chase gibt an, daß die Denkmalpflege nicht durch die Sammlung von Gebäuden oder die städtischen (Verwaltungs-)Strukturen erfolgreich verlief, sondern durch die Menschen. Charleston hat interessanterweise einen der wenigen historic districts in den USA, der über keine design guideline verfügt. Dies liegt vor allem am Alter und den administrativen Strukturen des historic district.

Obwohl der Charleston Old and Historic District einer starken Planungskomponente unterliegt, sind dennoch auch die privaten Wurzeln der Arbeit der HCF zu erwähnen. Diese Kombination konnte die heutige Form des historic district entstehen lassen. Die Hauptprobleme des historic district von Charleston liegen vor allem im Bereich des Tourismus, der im deutlichen Interessenkonflikt zu den Wünschen der Wohnbevölkerung steht. Ein weiteres Problem in Charleston stellt die Zonierung dar. Es gibt gegenwärtig Bestrebungen, weitere Industrie im historic district zuzulassen und auch aufgelockerte Gebäudevorschriften, die erlauben, alte Gebäude zu kleinen Wohneinheiten umzufunktionieren. Einige neighborhood-Organisationen haben schon deutliche Opposition angekündigt. Hervorzuheben ist, daß Charleston sich als eine der wenigen Städte aktiv für die niedrigen Einkommenschichten im Rahmen der Denkmalpflege einsetzt.

Als letztes sei noch die touristische Konkurrenz zu Savannah genannt. Es sollte jedoch bemerkt werden, daß es wesentlich sinnvoller wäre, den Tourismus gebündelt an die Region Charleston und Savannah zu binden, anstatt voneinander getrennt als regionale Konkurrenz aufzutreten.

Die Untersuchung in Charleston weist nach, wie eine denkmalpflegerische Verteidigung der historischer Gebäude mit allen existierenden Mängeln stattgefunden hat. Im Vergleich zu Society Hill und Beacon Hill sind in Charleston die planerischen Aktivitäten der Stadt in jüngerer Zeit sehr dominant. Die Interessenkonflikte bezüglich der touristischen Inwertsetzung durch die Stadt sind zwischen den Anwohnern und der Stadt ebenfalls noch nicht beigelegt. Der historic district in Charleston ist zwar gentrifiziert, weist jedoch auch heterogene, das heißt sozial und ethnisch vermischte Tendenzen auf. Die Initiative, Affordable Housing Projects in Society Hill oder Beacon Hill zu schaffen, konnte durch die entsprechenden neighborhood-Organisation in den 1970er Jahren erfolgreich abgewehrt werden.

Zusammenfassung: Savannah: Der Revolving Fund-Typ

Savannah hat, wie alle anderen Städte, die in den 1950er Jahren aufkommenden oder bereits existierenden Problemen US-amerikanischer Städte gehabt. Diese waren vor allem die einsetzende massive Suburbanisierung, die zunehmende Mobilität der Bevölkerung, der Ruf nach mehr Parkraum in den Städten und der Abzug der Manufakturen und anderer alter Industriezweige. Zum Teil wurde diesen Problemen in Savannah mit denkmalpflegerischen Programmen begegnet, zum Teil aber standen auch die Programme und Pläne der Denkmalpflege entgegen. Großprojekte bedingten den irreversiblen Abriß von historischer Bausubstanz. Die Einrichtung von mehr Parkraum ist im ursprünglichen Plan von Savannah nicht vorgesehen gewesen.

Ohne die beachtliche Pionierleistung der HSF, die dominierend und federführend von 1955 bis 1968 wirkte, wäre Savannah heute sicherlich nicht so gut erhalten. Bis 1968 wurden durch das Revolving Fund Program der HSF rund 130 Häuser restauriert. Private Initiative und Engagement haben das Interesse der Denkmalpflege in Savannah angeschoben und Abläufe mechanisiert. Nur die Erkenntnis der positiven push-Faktoren hat die lokale Regierung mobilisiert. Noch heute werden Richtlinienerweiterungen und historische oder architektonische Studien erst durch private Spenden ermöglicht. Erst die Zusammenarbeit von öffentlicher und privater Hand (Nutzung der Steuerabschreibungen) haben die weiträumige Erhaltung des historic district ermöglicht.

Hohe Gebäude mit größerer Rendite stehen der Idee entgegen, das historische Stadtbild zu erhalten. Es ist schwer zu bemessen, welchen ökonomischen Nutzen (Imagesteigerung, Touristenzahlen) eine historische Silhouette haben kann. In jedem Fall ist dies nur eine Rechnung, die nicht von Investorenseite, sondern nur von öffentlicher Seite her erstellt werden kann, da dies ein langfristiges Zusammenspiel einfordert. Die City of Savannah hat aus diesem Grund 1990 eine Studie in Auftrag gegeben, in der mittels eines Höhenindex Gebäudehöhen für alle Stadtgebiete festgelegt wurden.

Vor allem wird in Savannah deutlich, daß durch das organische Wachsen der neighborhood-Strukturen und durch die Abwesenheit großer Urban Renewal Projects die Erhaltung der historischen Bausubstanz möglich war. Die neighborhood-Erhaltungsmaßnahmen zeigen den Versuch auf, wie die Zahl der Umsiedlungen, die mit Maßnahmen der Denkmalpflege verbunden waren, minimiert werden konnten. Zunächst ist bewiesen, daß auch ohne die staatliche Lenkung die Erhaltung der historischen Bausubstanz möglich ist. Insbesondere ist im Victorian District die Denkmalpflege ohne weitere neighborhood-Strukturenverlagerungen, Umsiedlungen oder andere Veränderungsprozesse gelungen. Dieses Projekt unterlag vor allem gemeinnützigen Organisationen, die Ziele des Affordable Housing Program verfolgen. Obwohl Savannah in vielen Beziehungen eine ähnliche Struktur im Bereich der Genese der Denkmalpflege aufweist, läßt sich für Savannah eine geringere Projekt- und Planungsorganisation feststellen. Während in den beiden Städten Charleston und Savannah das Revolving Fund-System durch private Organisationen eingesetzt wurde, lenkte die Regierung in Charleston schneller in das Vorhaben ein.

Insbesondere in den letzten Jahren sind einige Strukturverbesserungen mittels denkmalpflegerischer Projekte zu verzeichnen: Der positive Effekt, den das Savannah College of Art and Design (SCAD) auf die Innenstadt hatte, ist für jeden Touristen unübersehbar: Durch die Anwesenheit der Studenten sind tagsüber neben den Touristen viele Menschen im Straßenbild präsent. Die Ausweitung der Programme wird diesen Effekt verstärken. Auch die revitalisierte Hafenfront mit der bewußt eingesetzten Funktionsmischung ist als gelungen einzustufen. Die Besucherzahlen, vor allem auch die große Anzahl an wiederkehrenden Besuchern aus der Region, belegen diese Einschätzung.

Zusammenfassung der Ergebnisse

Es wurde aufgezeigt, daß alle vier historic districts der untersuchten Städte einen essentiellen Beitrag zur urbanen Geschichte der USA leisten. Diese stadtpolitisch geplante und gesellschaftlich inszenierte Geschichte weist jedoch deutlich unterschiedliche Wege auf, die in dieser Studie mit den vier Erhaltungstypen (Urban Renewal-Typ in Philadelphia, Neighborhood-Typ in Boston, Planungstyp in Charleston, Revolving Fund-Typ in Savannah) klassifiziert wurden.

Die Formen der Erfolgskontrollen im Bereich praxisorientierter Stadtplanung sind in der Wissenschaft sehr umstritten. Es ist relativ einfach, den Status quo eines Landes oder einer Stadt zu untersuchen und zu bewerten, während Erfolgskontrollen die Dimensionen Vergangenheit und Gegenwert einbeziehen müssen. Im kombinierten Forschungsfeld Stadtplanung und Denkmalpflege spielen vielfältige, im empirischen Teil diskutierte Faktoren eine Rolle, so daß Erfolgskontrollen auf verschiedenen Bewertungsebenen unternommen werden müssen. Zum einen gibt es die baulich-funktionale und zum anderen die gesellschaftlich-historische Ebene. Beide Bewertungsskalen sind unterschiedlich und kommen zum Teil bei solchen Erfolgsmessungen zu unterschiedlichen Ergebnissen.

Städtische Denkmalpflege in den USA: Ein nationaler Vergleich

Denkmalpflege ist heute innerhalb der Gesellschaft weitgehend akzeptiert. Methoden und Gelder werden durch private und öffentliche Institutionen, wie der empirische Teil gezeigt hat, multifunktional eingesetzt.

In den USA gab es lange Zeit keine festgeschriebenen baulichen Richtlinien und auch kein überregionales Raumordnungsprogramm. Dieser Zustand förderte die Wildwuchsarchitektur der amerikanischen Städte erheblich. Wildwuchsarchitektur, Suburbanisierung, ökonomische Talfahrt der Städte und zunehmende Negierung der Städte haben einige Bürger auf die zunehmenden Mängel und planerischen Defizite der Städte aufmerksam gemacht. Die regionale Planung und die Frage, wer der Nachbar ist, geht das Individuum unmittelbar etwas an. Vor allem seit den 1970er Jahren nimmt der Einzelbürger an der Gestaltung der eigenen Umgebung in der Regel aktiv teil und ergreift Initiative. Der Denkmalschutz ist in diesem Zusammenhang ein Medium, die Homogenisierung der neighborhood voranzutreiben. Es ist auffällig, wie emotional und persönlich die US-Amerikaner einige Veränderungen aufnehmen. Planungs- und Bauvorhaben sind in den USA weitgehend eine lokale Angelegenheit der Bürgerschaft. Ein klein- oder großmaßstäbiges Konzept der Raumordnung und demnach auch der denkmalpflegerischen Erhaltung existiert nicht beziehungsweise nur rudimentär.

Die vier historic districts der Städte Philadelphia, Boston, Charleston und Savannah weisen in mehreren Beziehungen ähnliche Strukturen und Organisationsformen sowie vergleichbare Probleme und Potentiale auf. Die Anfänge der Denkmalpflege-Bewegung ist in fast allen Städten auf den Einsatz von Einzelpersonen für Einzelobjekte zurückzuführen.

Alle Städte mussten in den 1950er und 1960er Jahren mit ähnlichen Problemen kämpfen. Zu diesen zählten hauptsächlich: Niedergang der lokalen Industrie (damit Arbeitsplatzmangel und ökonomischer Engpass), Suburbanisierung, Akzeptanzbildung innerhalb der Bevölkerung, Geschichte und Kulturdefinition (heritage-Bewusstseinsbildung), Parkplatzsituation, Mikroarrangements in den Städten, wie zum Beispiel der Bau einer Autobahn oder eines Kongreßzentrums.

In den vier Städten entwickelte sich dann aus diesen Problemen eine denkmalpflegerische Eigendynamik, die dazu führte, dass sich die einzelnen Städte unterschiedlich entwickel(te)n.

Die untersuchten historic districts sind, die touristisch am meisten frequentierten Gebiete aller Städte. Direkt angrenzende Gebiete und neighborhoods sind häufig noch immer heruntergekommen. Bei der internationalen ebenso wie der nationalen Beurteilung ist weiterhin zu beachten, ob sich die Faktoren voneinander isolieren lassen oder ob nur eine ganzheitliche Betrachtung angemessen ist. Vergleiche sollten sich zunächst auf die essentiellen Faktoren der Länder beschränken.

Die geförderte Denkmalpflege in den USA stößt deutlich an ihre Grenzen. Es gibt keine übergeordneten Klassifikationssysteme, welche die Bestände der schützenswerten amerikanischen Gebäude systematisch und überregional aufgenommen haben. Wegen der relativ schlechten Dokumentation kommt noch hinzu, daß zum Teil unbekannt ist, ob die jeweiligen als historisch deklarierten Gebäude überhaupt noch existieren und in welchem Zustand sie sich befinden. Es obliegt demzufolge den einzelnen lokalen Regierungen, eine effiziente lokale Denkmalpflege nach eigenem Ermessen und eigenen Maßstäben zu betreiben. Statistische Aufnahmen sind demnach deutlich der regionalen Obhut zugeordnet.

Nachteile der praktizierten Denkmalpflege in den untersuchten historic districts sind, daß Gebiete, die schon immer benachteiligt waren, auch weiterhin bei der denkmalpflegerischen Inwertsetzung nur zweitrangig behandelt werden. Erfolgreiche Lösungsstrategien aus diesem Dilemma konnte für Savannah und Boston mit den ‚Affordable Housing'-Projekten nachgewiesen werden. Eine ästhetische architektonische Prüfung konnte hier gewährleisten, daß ein erwünschter gesellschaftlich-historischer Charakter der neighborhoods bestehen bleibt.

Die Opposition durch African-Americans gegen die Einrichtung beziehungsweise Erweiterung von historic districts ist ebenfalls erwähnt worden. Offenbar sind die historic districts ein weiteres städtisches Forum, in dem ethnische Differenzen und Minoritätenprobleme offensichtlich werden. Viele Minoritätengruppen verfügen über keine oder keine hinreichend einflussreiche Lobby, die ihre Bedürfnisse artikuliert.

Nicht zu übersehen sind in diesem Zusammenhang die Auswirkungen des neueren Geschichtsverständnisses und die Betonung von heritage-Aspekten im Rahmen denkmalpflegerischer Aktivitäten in den USA. Denkmalpflege ist hier auch als Zähmung und Kultivierung der vielen globalen, kulturellen und lokalen Einflüsse und Kräfte zu verstehen, die auf die Städte einwirken. Dennoch können auch die historic districts stadtplanerisch so eingesetzt werden, daß es zu einer

Verflechtung von Räumen kommt. Diese Räume können durch einen Stadtbezirk, ein County, die SMSA der Stadt oder eine ganz Region repräsentiert werden. Im Süden ist neben den ähnlichen Strukturen wie im Norden noch zusätzlich die Katastrophenprävention wichtig.

Es steht heute außer Frage, daß Denkmalpflege den Tourismus einer Stadt oder einer Region ankurbelt. Die unterschiedlichen und besonderen Charaktere von Architektur und Orten ziehen regionale, nationale und internationale Touristen an. Dies gilt für alle vier untersuchten historic districts. Es werden ebenso neighborhood-Konditionen und deren Außenbilder verbessert. Kongresse und Festivalausrichtungen weiten das Angebot für den Tourismus maßgeblich weiter aus. Mehrere Beispiele wurden hierzu im empirischen Teil benannt. Die touristischen Ziel- und Interessenkonflikte, die dabei entstanden, sind für Charleston exemplarisch herausgearbeitet worden. Es wäre wünschenswert, daß sich die Denkmalpflege in einigen Städten der USA nicht nur auf die touristische Attraktionen beschränken würde. Durch den angewandten Methodenpluralismus dieser Studie sind die vielseitigen Verflechtungen der Denkmalpflege aufgezeigt. Gegenwärtig handelt es sich um eine zyklische Form der Denkmalpflege, sie ist tagespolitisch aktuell.

Die Gründe dafür, Denkmalpflege zu betreiben, variieren in den Städten jedoch erheblich. Übereinstimmung finden die Städte in der Meinung, daß die Denkmalpflege „.. the meaning of place and community identity ..“ (Hodder 1999, S.439) vereint. Ebenso wird die bauliche Substanz auch im Sinne der Schule des Urban Design nachhaltig geschützt.

Es ist weiterhin zu begrüßen, daß viele Gemeinden denkmalpflegerische Ziele zu ihren eigenen machen. Die grass-root-Maximen werden dabei beachtet und die Denkmalpflege erfährt gleichzeitig einen breiteren offiziellen Charakter mit der Vertretung öffentlicher Interessen (public welfare), für die sogar enteignet werden darf. Der Prozeß der Verdrängung sozial schwacher Bevölkerungsteile (Gentrification) ist, obwohl es schon oft erkannt und quantifiziert wurde, weiterhin eine der schwerwiegensten Nebenwirkungen, die bei der Errichtung von historic districts in Erscheinung tritt. An dieser Stelle sind die vorgestellten exklusivierenden Tendenzen in Beacon Hill und Society Hill besonders hervorzuheben. Interessenkonflikte bezüglich der Authentizität und der allgemein städtebaulichen Positionierung werden dabei hervorgehoben. Der einseitigen Ressourcennutzung (Fördermittel, Fond-Nutzungen) wird durch Sonderprogramme von staatlicher Seite her entgegengewirkt.

Die Unterschiede der denkmalpflegerischen Verfahrensweisen machen deutlich, daß nur die Methoden der „individuellen Stadtgeographie“ (Denecke 1989) zeigen können, welche Projekte Chancen haben können. Einfache statistische Analysen können zu Fehlschlüssen führen. Nur eine interdisziplinäre Untersuchung kann mit Hilfe des Methodenpluralismus zu einer ganzheitlichen Bewertung kommen, die dann für weitere Neueinrichtungen von historic districts oder für das Umfeld der historic districts von Nutzen sind.

Schlußbetrachtung

Die Formen und Umsetzungen der Denkmalpflege in den historic districts der Städte Philadelphia, Boston, Charleston und Savannah wurden exemplarisch und theoretisch diskutiert. Es wurde dabei die Mehrdimensionalität des Konzeptes der historic districts hervorgehoben. Die Untersuchungen der vorliegenden Dissertation haben dadurch gezeigt, wie vielfältig die einzelnen kulturhistorischen Entstehungsgründe, Methoden und Motivationen beim Konzept der historic districts sind. Das Konzept der historic districts hat sich seit seinen Anfängen (Charleston im Jahr 1931) nicht nur planerisch-konzeptionell verändert, es hat vielmehr auch aufgrund historischer Einbeziehung von ethnischen Minderheiten eine breitere gesellschaftliche Akzeptanzbasis innerhalb der US-amerikanischen Gesellschaft erhalten.

Auf der thematisch-sachlichen Ebene können die folgenden Untersuchungsergebnisse festgehalten werden: Es wurde auf den Verlust der Niedrigmieten in den historic districts durch den massiven Gentrificationprozeß hingewiesen. Einlenkend gibt es gegenwärtig jedoch neuere staatliche Projekte, diese gesellschaftlichen Disparitätenbildung abzuschwächen. Dabei entstehen grundsätzliche Kontroversen durch die Definition über das „Wohl der Gesellschaft“: Es kann staatlicherseits eingelenkt werden oder es kann gemäß den Maximen des Privatkapitalismus eine staatlich-planerische Zurückhaltung erfolgen. Planungen mit integrativem Charakter stehen den segregativen Tendenzen der US-amerikanischen Gesellschaft insbesondere in städtischen historic districts entgegen. Gerade dort ist dieses Problem wegen einer traditionalistische Haltung gegenüber historisch-kulturellen Werten und der Sicherung der quality-of-life-issues festzustellen. Obwohl unterschiedliche prozessuale Mikrostrukturen in den historic districts existieren, sind in allen deutlich exklusive Haltungen nachgewiesen worden. Lediglich der Savannah Historic District weist die beschriebenen Merkmale mit dieser Deutlichkeit nicht auf.

Auf der methodischen Ebene lassen sich die folgenden Ergebnisse zusammenfassen: Stadterhaltung darf nicht Stagnation heißen, das heißt, eine Anpassung an moderne gesellschaftliche Konditionen ist notwendig. Räumliche Vitalität muß weiterhin den städtischen Raum prägen. Es sollte dabei auch nicht vergessen werden, daß die Neubauten von heute wiederum als Denkmäler von morgen ausgewählt werden können. Moderne Denkmalpflegebewegungen engagieren sich auch gegen die zunehmende gesellschaftliche Anonymität oder zu schnelle Bauvorhaben durch Investoren in den neighborhoods („lokale Innovation“).

Stadtbildpflege durch die Denkmalpflege in den historic districts ist mehr als nur Fassadenerneuerung oder die Förderung von neuen Investitionen. Die Förderung von Anregungen, Orientierungs- und sozialen Bezugspunkten im städtischen Raum oder Quartier ist ebenso von großer Bedeutung. Die Imagepflege einer Stadt sollte von der Fremdenverkehrsorientierung auf die Bevölkerung ausgeweitet werden. Problematisch ist bei diesem Ansatz, daß dabei auch irrationale Entscheidungsmomente von Neubürgern und Investoren unvorhersehbar in Erscheinung treten können.

Eines der wirksamsten Mittel, Denkmalpflege zu betreiben, bleibt sicherlich die Überführung in öffentlichen Besitz. Jedoch kann nicht negiert werden, daß das Instrument des historic district im US-amerikanischen Privatkapitalismus mit den zugehörigen lokalen Planungsrechten und -instrumenten eine weitere erfolgreiche Methode darstellt, historische Bauwerke zu erhalten. In diesem Zusammenhang sind die neighborhood-Organisationen in den USA die dominanten und maßgeblichen Akteure.

Es konnte nachgewiesen werden, daß die historic districts heute als vielseitiges und sinnvolles Stadtplanungsinstrument eingesetzt werden (können). Historic districts sind ein Weg, historische Substanz zu schützen und gleichzeitig aufzuwerten und zu nutzen. Die Interdisziplinarität hat gewährleistet, daß sich für die verschiedenen wissenschaftlichen Felder, die ebenfalls im Forschungsfeld des Denkmalschutzes arbeiten, vielseitige Informationen ableiten lassen. Diese Verständigung ist wichtig, weil dadurch die städtische Denkmalpflege betreffenden Schlüsselinformationen ausgetauscht werden (demographische Informationen, Segregationstendenzen, Vitalität der Städte, Stadtimage, Außen- und Innenbild der Städte).

Der Vergleich und die Übertragung der Ergebnisse auf Deutschland hat schließlich gezeigt, daß politische, ökonomische und sozio-kulturelle Faktoren ganzheitlich mitberücksichtigt werden müssen und somit einen prozessualen Vergleich nur bedingt zulassen. Eine deckungsgleiche Übertragung auf deutsche Verhältnisse ist darum nur punktuell möglich, wie beispielsweise die Ausführungen über das Konzept der „living history“ gezeigt haben.

Bibliography

The only titles included here are those which appear several times in the text or which have a direct connection with the topic of this study.

Abel, Chris (1997): Architecture and Identity: Towards a Global Eco-Culture. Oxford University Press. Oxford.

Abler, Ronald (1976): A Comparative Atlas of America's Great Cities. Twenty Metropolitan Regions. Minnesota.

ACHP – Advisory Council on Historic Preservation (ed.) (1988): President's Historic Preservation Awards. Washington, DC.

ACHP – Advisory Council on Historic Preservation (ed.) (1992): President's Historic Preservation Awards. Washington, DC.

ACHP – Advisory Council on Historic Preservation (ed.) (1993): National Historic Preservation Act of 1966, as Amended. Washington, DC.

ACHP – Advisory Council on Historic Preservation (ed.) (1998): Report to the President and Congress 1996–1997. Washington, DC.

Adams, Carolyn T. (1990): Philadelphia – The Private City in the Post-Industrial Era. In: Bernard, Richard M. (ed.): Snowbelt Cities: Metropolitan Politics in the Northeast and Midwest Since World War II. Bloomington.

Adams, Carolyn T. et al. (1991): Philadelphia: Neighborhoods, Division and Conflict in a Postindustrial City. Temple University Press. Philadelphia.

Adams, John S. (ed.) (1978): Contemporary Metropolitan America. Cities of the Nation's Historic Metropolitan Core. Cambridge, Massachusetts.

Adams, Willi Paul (ed.) (1976): Die Vereinigten Staaten von Amerika. Kartenteil. Campus-Verlag Frankfurt/New York.

AIA – American Institute of Architects (ed.) (1965): Urban Design: The Architecture of Towns and Cities. New York.

AIA – American Institute of Architects (ed.) (1996): The Philadelphian Architect. Philadelphia.

AIA – American Institute of Architects (ed.) (1997): Readings in Historic Preservation. Why? What? How? Washington, DC.

Alba, Richard D. (1990): Ethnic Identity: The Transformation of White America. Yale University Press. New Haven.

Albers, Gerd (^{2}1992): Stadtplanung: Eine praxisorientierte Einführung. Darmstadt.

Albrecht, Wilma R. (1989): New York - Phoenix in der Asche? In: Die Alte Stadt. No. 4/1989. pp. 607–609.

Alexander, Christopher (ed.) (1987): A New Theory of Urban Design. Oxford University Press. New York, London.

Alisch, Monika (ed.) (1998): Stadtteilmanagement. Voraussetzungen und Chancen für die soziale Stadt. Verlag Leske + Budrich. Opladen.

Alotta, Robert I. (1990): Mermaids, Monasteries, Cherokees and Custer: The Story Behind Philadelphia's Street Names. Chicago.

Andriot, John L. (ed.) (1979): Township Atlas of the United States. McLean, Virginia.

APA – American Planning Association (1992): Innovative Tools for Historic Preservation. Washington, DC.

Architectural Record (1976): The Home Towns Come Back. Savannah. Rehabilitating Houses is Nothing New in Savannah. But Rehabilitating Victorian Houses for Low-Income Families Is – And It's Working. Savannah.

Assmann, Jan (1999): Das kulturelle Gedächtnis. Becksche Reihe. Nördlingen.

Atteslander, Peter/Hamm, Bernd (ed.) (1974): Materialien zur Siedlungssoziologie. Köln.

Attoe, Wayne/Down, Logan (^{2}1992): American Urban Architecture. Catalyst in the Design of Cities. University of Carlifornia Press. Berkeley.

Bach, Penny Balkin (1992): Public Art in Philadelphia. Philadelphia.

Bacon, Edmund (^{3}1976a): A Case Study in Urban Design. In: Journal of the American Institute of Planners. Vol. 26, No. 3/1960, pp. 224–235.

Bacon, Edmund (^{3}1976b): Design of Cities. New York.

Bacow, Adele Fleet (1995): Designing the City: A Guide for Advocates and Public Officials. Island Press. Washington, DC.

Baker, David (1983): Living With the Past: The Historic Environment. Bletsoe, UK.

Balack, Ulf (2000): Die *white collar* Gruppe in Philadelphia: Entwicklung, Struktur und Mobilität, 1900–1970. Dissertation. Universität Hamburg.

The Balch Institute for Ethnic Studies (ed.) (1992): Selections from the Museum Collections. Philadelphia.

Baltzell, Digby E. (1958): Philadelphia Gentlemen. The Making of a National Upper Class. Free Press. Glencoe, Illinois.

Baltzell, Digby E. (1980): Puritan Boston and Quaker Philadelphia. Free Press. Philadelphia.

Barrington, Lewis (1941): Historic Restaurations of the Daughters of the American Revolution. New York.

Bathel, Diane L. (1996): Historic Preservation: Collective Memory and Historical Identity. Rutgers University Press. New Brunswick.

Baumann, John F. (1987): Public Housing, Race and Renewal. Urban Planning in Philadelphia 1920–1974. Philadelphia.

BBR – Bundesforschungsanstalt für Landeskunde und Raumordnung (ed.) (1991): Frauen und räumliche Planung. In: Materialien zur Raumentwicklung Heft 38. Bonn.

BBR – Bundesministerium für Raumordnung, Bauwesen und Städtebau (ed.) (1994): Standortkultur und Städtebaulicher Denkmalschutz. In: Informationsdienste Städtebaulicher Denkmalschutz No. 11. Berlin, Bonn.

BBR – Bundesministerium für Raumordnung, Bauwesen und Städtebau (ed.) (1996a): Kommunales Management zur Stadterneuerung in denkmalwerten Städten. Dokumentation der Fachtagung in Quedlinburg. In: Informationsdienste Städtebaulicher Denkmalschutz No. 17/18. Berlin, Bonn.

BBR – Bundesministerium für Raumordnung, Bauwesen und Städtebau (ed.) (1996b): Neues Bauen im historischem Bestand der Altstädte. Dokumentation der Fachtagung in Leipzig. In: Informationsdienste Städtebaulicher Denkmalschutz No. 19. Berlin, Bonn.

BBR – Bundesministerium für Verkehr, Bau- und Wohnungswesen (ed.) (1998): Städtebaulicher Denkmalschutz in den neuen Ländern. In: Informationsdienste Städtebaulicher Denkmalschutz No. 22. Berlin, Bonn.

BBR – Bundesamt für Bauwesen und Raumordnung (ed.) (1999): Perspektiven für die Region als Planungs- und Handlungsebene. Heft 9–10/1999. Bonn.

Beatly, Timothy (1994): Ethical Land Use. Principals of Policy and Planning. John Hopkins University Press. Baltimore.

Beauregard, Robert A. (1989a): City Profile: Philadelphia. In: Cities. No. 6/1989. pp. 300–308.

Beauregard, Robert A. (ed.) (1989b): Atop the Urban Hierarchy. Totowa, New York.

Beauregard, Robert A. (1993): The Turbulence of Housing Markets: Investment, Disinvestment and Reinvestment in Philadelphia, 1963–1986. In: Knox, Paul L. (ed.): The Restless Urban Landscape. Englewood. pp. 55–82.

Becker, Franz J./Schrand, Hermann (1978): Stadtgeographie. Düsseldorf.

Bednar, Michael J. (1989): Interior Pedestrian Spaces. New York.

Behrens, Jan (1995): Zur Rolle der „neighborhood" in der US-amerikanischen Großstadt – mit einer Analyse zweier Stadtdistrikte in Philadelphia. Unpublished *Diplomarbeit*. Universität Hamburg.

Bergen, Philip (1990): Old Boston in Early Photographs, 1850–1918. Prints from the Bostonian Society. Dover Publications. New York.

Berkeley, Ellen Perry/McQuaid Matilda (eds.) (1989): Architecture: A Place for Women. Smithsonian Institute Press. Washington, DC.

Berry, Jeffrey, M. et al. (1993): The Rebirth of Urban Democracy. Washington, DC.

BHCA – Beacon Hill Civic Association (1975): Beacon Hill Architectural Handbook. Guidelines for Preservation & Modification. Boston.

BHCA – Beacon Hill Civic Association (1999): Beacon Hill News Letter. No. 5/1999. Boston.

Birch, Eugenie Ladner/Roby, Douglas (1984): The Planner and the Preservationist: An Uneasy Alliance. In: Journal of the American Planning Association. Vol. 50. No. 2/1984. pp. 194–207.

Bissinger, Buzz (1997): A Prayer for the City. Vintage Books, Random House. New York.

Bland, Sidney (1994): Preserving Charleston's Past, Shaping its Future. Boston.

BLC – Boston Landmark Commission (1975): An Act Establishing the Boston Landmarks Commission. Boston.

BLC – Boston Landmarks Commission (1979a): Central Business District Preservation Study. Boston.

BLC – Boston Landmarks Commission (1979b): Building and Streetscape Preservation Survey for Boston's Theater District. Boston.

BLC – Boston Landmarks Commission (1980a): Central Business District Preservation Study. Part II-Summary of Findings. Boston.

BLC – Boston Landmarks Commission (1980b): Central Business District Preservation Study. Part II – Summary of Findings. Boston.

BLC – Boston Landmarks Commission (1982): 1981 Survey and Planning Grant. Part I. South Boston. Boston.

Blockson, Charles L. (^{2}1995): The Underground Railroad. New York.

Bloomer, Kent C./Moore, Charles W. (1975): Body, Memory and Architecture. Yale University Press. New Haven.

Bluestone, Barry/Harrisson, Bennett (1982): The Deindustrialization of America: Plant Closings, Community, Abandonment and the Dismanteling of Basic Industry. New York.

Blume, Helmut (^{2}1979): USA. Die Regionen der USA. Wissenschaftliche Länderkunden Band 9/II. Darmstadt.

Blume, Helmut ([3]1987): USA. Der Großraum im strukturellem Wandel. Wissenschaftliche Länderkunden Band 9/I. Darmstadt.

Born, Karl Martin (1995): Raumwirksames Handeln von Verwaltungen, Vereinen und Landschaftsarchitekten zur Erhaltung der Historischen Kulturlandschaft und ihrer Einzelelemente. Eine vergleichende Untersuchung in den nordöstlichen USA (New England) und der Bundesrepublik Deutschland. Dissertation Universität Göttingen.

Bornheim, Werner (1956): Zur Denkmalpflege in Nordamerika. In: Deutsche Kunst- und Denkmalpflege. Jg. 1956. München, Berlin. pp. 95–101.

Boston Edison/Historic Boston Incorporated/Partners for Sacred Places/National Trust for Historic Preservation (eds.) (1990): Sacred Trusts. Reducing Energy Costs in Religious Buildings. A Workbook for Congregation Leaders. Boston.

Bostonian Society (1700–1999): File Beacon Hill. Boston.

Bourne, Larry S. (1981): The Geography of Housing. In: Scripta Series in Geography. London, UK.

Boyer, Christine M. (1983): Dreaming the Rational City: The Myth of American City Planning. MIT Press. Cambridge.

Boyer, Christine M. ([3]1996): The City of Collective Memory. Its Historical Imaginary and Architectural Entertainments. MIT Press. Cambridge.

BRA – Boston Redevelopment Authority (ed.) (1981): The Boston Sign Code. Boston.

BRA – Boston Redevelopment Authority (ed.) (1995a): Facts About Boston Neighborhoods. Boston.

BRA – Boston Redevelopment Authority (1995b): Ways to Look at Boston. Boston.

Brant, Steward (1995): How Buildings Learn: What Happens After They Are Built. Penguin Press. New York.

Brownlee, David B. (1989): Building the City Beautiful. Lunenburg, Vermont.

Buchholz, Hans Henning (1997): Städtebauliche Denkmalpflege in Klein- und Mittelstädten der alten Bundesrepublik. In: Schriftenreihe zur Kulturwissenschaft Band 12. Hamburg.

Bundesministerium für Verkehr, Bau- und Wohnungswesen (ed.) (1999): Wirkungsanalyse Städtebaulicher Denkmalschutz in den neuen Ländern. In: Informationsdienste Städtebaulicher Denkmalschutz No. 23. Berlin, Bonn.

Burkart, Günter (1993): Individualisierung und Elternschaft – Das Beispiel USA. In: Zeitschrift für Soziologie. Jg. 22, Heft 3, No. 4/1993. pp. 159–177.

Cantor, George (1981): Historic Landmarks of Black America. Detroit.

Campbell, Robert (1995): The Preservation Battle. In: National Trust for Historic Preservation (ed.): Historic Preservation Forum. No. 4/1995. Washington, DC. pp. 18–24.

Campbell, Scott/Fainstein, Susan (1996): Readings in Urban Theory. Cambridge, Massachusetts 1996.

Campbell, Scott/Fainstein Susan ([2]1997): Readings in Planning Theory. Blackwell Publisher. Malden, Massachusetts 1997.

Catanese, Anthony James/Snyder, James (eds.) (1979): Introduction to Urban Planning. New York.

Catanese, Anthony James (1984): The Seven Golden Rules: Politics in Planning and Urban Design. In: APA – American Planning Association (ed.): Urban Design Review. Vol. 7, No. 4/1994. pp. 11–12.

CCD – Center City District (ed.) (1994): Center City Design: Design Standards of the Center City District. Philadelphia.

Chadbourne, Christopher (1993): A City of Streets and Squares. In: National Trust for Historic Preservation (ed.): Historic Preservation Forum. No. 9–10/1993. Washington, DC. pp. 37–43.

Chapin, Stuard, F./Kaiser, Edward J. (1979): Urban Land Use Planning. University Press of Urbana. Urbana, Illinois.

Chatham County Savannah Metropolitan Planning Authority (1969): Ideas for Downtown. Savannah.

Chiat, Marilyn J. (1997): America's Religious Architecture: Sacred Places for Every Community. Preservation Press. New York.

City of Boston (ed.) (1975): Background Information, Planning Issues and Preliminary Neighborhood Improvement Strategies. Boston.

City of Boston (ed.) (1997): Historic Beacon Hill District. Architectural Guidelines. Boston.

City of Charleston/Board of Architectural Review (eds.) (1996): District Map Charleston. Charleston.

City of Charleston (ed.) (1996a): Submittal and Approval Information (About Historic Preservation in Charleston, SC). Charleston.

City of Charleston (ed.) (1996b): Charleston Code. Charleston.

City of Charleston (ed.) (1996c): Historic Preservation. Article 2: Land Use Regulation. Charleston.

City of Charleston (ed.) (1996d): Historic Preservation. Article 3: Site Regulation. Charleston.

City of Charleston (ed.) (1996e): Commercial District Signs. Charleston.

City of Philadelphia (ed.) (1962): Community Renewal Program. Technical Report No. 1. Philadelphia.

City of Philadelphia (ed.) (1966): The Negro Housing: A Program for Philadelphia. Technical Report No. 18. Philadelphia.

City of Philadelphia (ed.) (1984): An Ordinance. Bill No. 318. Philadelphia.

City of Philadelphia (ed.) (1993): Zoning Ordinance of the City of Philadelphia. As Amended, August 1st, 1993. Philadelphia.

City of Philadelphia (ed.) (1996): City of Philadelphia: Five-Year Financial Plan. Economic Development 1997–2000. Philadelphia.

City of Philadelphia (ed.) (1999): Philadelphia. A New Urban Direction. St. Joseph's University Press. Philadelphia.

City of Savannah/Metropolitan Planning Commission Chatham County (1997a): Manual for Development in the Savannah Historic District. Savannah.

City of Savannah/Metropolitan Planning Commission Chatham County (1997b): Historic District Zoning Ordinance (Savannah). Savannah.

Clark, David (^{2}1988): Post – Industrial America: A Geographical Perspective. New York.

Clark, David (1996): Urban World/Global City. Routledge Publisher. New York.

Conzen, Michael P./Carr, Kay J. (eds.) (1988): The Illinois & Michigan Canal National Heritage Corridor: A Guide to its History and Sources. DeKalb, Illinois.

Conzen, Michael P./Lewis, George K. (eds.) (1976): Boston: A Geographical Portrait. In: Adams, John S. (ed.): Contemporary Metropolitan America. Cities of the Nation's Historic Metropolitan Core. Cambridge, Massachusetts. pp. 51–136.

Costonis, John J. (1974): Space Adrift. Saving Urban Landmarks Through the Chicago Plan. University of Illinois Press. Chicago.

Costonis, John J. (1989): Icons and Aliens. University of Illinois Press. Chicago.

Council of Economic Advisors (ed.) (1998): Changing America: Indicators of Social and Economic Well-Being by Race and Hispanic Origin. Washington, DC.

Cohen, Michael A. et al. (eds.) (1996): Preparing for the Urban Future: Global Pressuring, Local Forces. Baltimore.

Cohen, Nahoum (1999): Urban Conservation. MIT Press. Cambridge, Massachusetts.

Copps, David H. (1995): Views from the Road. Island Press. Washington, DC.

Cotter, John L. et al. (²1994): The Buried Past. An Archeological History of Philadelphia. University of Pennsylvania Press. Philadelphia.

Croci, Giorigio (1998): The Conservation and Structural Restoration of Architectural Heritage. Southampton, UK.

Cullingworth, Barry (1997): Planning in the USA. Policies, Issues and Processes. London, New York.

Cuttler, William III (1980): The Persistent Dualism. Centralisation and Decentralisation in Philadelphia 1854–1975. In: Cuttler, William III (ed.): The Divided Metropolis: Social and Spatial Dimensions of Philadelphia. Westport.

Curdes, Gerhard (1993): Stadtstruktur und Stadtgestaltung. Berlin, Köln.

Cybriwsky, Roman A. (1986): The Fashioning of Gentrification in Philadelphia. In: Urban Resources. No. 3/1986. pp. 27–53.

Cybriwsky, Roman A. (1992): Revitalization Trends in Downtown–Area Neighborhoods. In: Brunn, Stanley D. (ed.): The American Metropolitan System: Present and Future. New York.

Cybriwsky, Roman A. et al. (1986): The Political and Social Construction of Revitalized Neighborhoods: Society Hill, Philadelphia, and False Creek, Vancouver. In: Smith, Neil/Williams, Peter (eds.): Gentrification of the City. Winchester, Massachusetts. pp. 92–120.

Cybriwsky, Roman A./Western, John (1982): Revitalizing Downtown: By Whom for Whom? In: Geography and the Urban Environment. No. 5/1982. Chinchester. pp. 343–366.

Dahms, Hellmuth Günther (1983): Grundzüge der Geschichte der Vereinigten Staaten. Wissenschaftliche Buchgesellschaft Band 21. Darmstadt.

Dangschat, Jens S./Blasius, Jörg (eds.) (1990): Grundlagen und Folgen. In: Gentrification: Die Aufwertung innenstadtnaher Wohngebiete. Campus Verlag. Frankfurt am Main. pp. 11–35.

Datel, Robin Elisabeth (1985): Preservation and a Sence of Orientation for American Cities. In: The Geographical Review 1985. pp. 125–141.

Datel, Robin Elisabeth (1990): Southern Regionalism and Historic Preservation in Charleston, South Carolina, 1920–1940. In: Journal of Historical Geography. Vol. 16, No. 2/1990. pp. 197–215.

Datel, Robin Elisabeth/Dingemans, Denis J. (1980): Historic Preservation and Urban Change. In: Urban Geography. Vol. 1, No. 3/1980. pp. 229–253.

Datel, Robin Elisabeth/Dingemans, Denis J. (1988): Why Places are Preserved: Historic Districts in American and European Cities. In: Urban Geography. Vol. 9, No. 1/1988. pp. 37–52.

Davies, Jonette Ella (1997): Roles of Historic Theatres in Urban Revitalization: The Detroit Theatre District and Philadelphia's Avenue of the Arts. Unpublished M.A. thesis in the Historic Preservation Department at the University of Pennsylvania. Philadelphia.

Davis, Allen F./Haller, Mark H. (1973): The Peoples of Philadelphia. Temple University Press. Philadelphia.

Davis, Allen F. et al. (1983): Still Philadelphia: A Photograph History 1890–1940. Philadelphia.

Delafons, John (1997): Politics and Preservation. Surrey, UK.

Delaware Valley Archivists Group (ed.) (1992): Directory of Archival and Manuscript Repositories in Southeastern Pennsylvania. Philadelphia.

Denecke, Dietrich (1989): Stadtgeographie als geographische Gesamtdarstellung und komplexe geographische Analyse einer Stadt. In: Die Alte Stadt. No. 1/1989. pp. 3–23.

Denton, Nancy A./Massey, Douglas S. ([7]1996): American Apartheid: Segregation and the Making of the Underclass. Cambridge, Massachusetts.

Deutsches Nationalkomitee für Denkmalschutz (ed.) (1990): Architektur und Städtebau der 50er Jahre. Schriftenreihe Band 41. Paderborn.

Deutsches Nationalkomitee für Denkmalschutz (ed.) (1994): Denkmalpflege als Standort- und Wirtschaftsfaktor. Schriftenreihe Band 49. Paderborn.

Deutsches Nationalkomitee für Denkmalschutz (ed.) (1997): Kirchen in Not. Schriftenreihe Band 56. Paderborn.

Deutsches Nationalkomitee für Denkmalschutz (ed.) ([2]1998): Rekonstruktion in der Denkmalpflege. Überlegungen, Definitionen, Erfahrungsberichte. Schriftenreihe Band 57. Paderborn.

Deutsches Nationalkomitee für Denkmalschutz (ed.) ([2]1999a): Kursbuch Denkmalschutz. Schriftenreihe Band 50. Paderborn.

Deutsches Nationalkomitee für Denkmalschutz (ed.) ([2]1999b): Auf dem Weg ins 21. Jahrhundert-Denkmalschutz und Denkmalpflege in Deutschland. Schriftenreihe Band 61. Paderborn.

Dobby, Alan (1978): Conservation and Planning. London.

Dochter, Richard/Wagner, Rich (1991): Brewerytown, USA. In: Pennsylvania Heritage. No. 3/1991. Harrisburg. pp. 24–31.

DVRPC – Delaware Valley Regional Planning Commission (ed.) (1985): Regional Infrastructure Evaluation and Analysis for Southeastern Pennsylvania. Summary Report. Philadelphia.

DVRPC – Delaware Valley Regional Planning Commission (ed.) (1993): Rating the Region. Direction 2020. Philadelphia.

Dyroff, Hans-Dieter (1997): Welterbe und Stadtentwicklung: Das Beispiel Potsdam. Bonn.

Eberlein, Harold D. (1960): Philadelphia Houses a Proud Past. In: National Geographic. Vol. 118, No. 8/1960. pp. 151–192.

Ellin, Nan (1996): Postmodern Urbanism. Blackwell Publisher. Oxford, UK.

Ershkowitz, Miriam/Zikmund, Joseph II (1973): Black Politics in Philadelphia. Basic Book Inc. New York.

Falke, Andreas (1992): Regional- und Stadtentwicklung. In: Die Vereinigten Staaten von Amerika. Band 1. Frankfurt.

Featherstone, Kevin/Ginsberg, Roy H. (1996): The United States and the European Union in the 1990s. Basingstoke, Hampshire.

Feilden, Bernhard M. ([5]1997): Conservation of Historic Buildings. Architectural Press. Oxford, UK.

Finkel, Kenneth/Oyama, Susan (1988): Philadelphia Then and Now. New York.

Finkel, Kenneth (ed.) (1994): Philadelphia Almanac and Citizen's Manual. Philadelphia.

Fisher, Robert (1994): Let the People Decide: Neighborhood Organizing in America. Social Movements Past and Present. New York.

Fitch, James Marston ([3]1995): Historic Preservation: Curatoral Management of the Built World. Charlottesville.

Fleming, John et al. (eds.) ([4]1991): Dictionary of Architecture. Penguin Press. London.

Fletcher Hodges III (1987): Two Viewpoints on the Religious Properties Crisis. In: National Trust for Historic Preservation (ed.): Preservation Forum. No. 4/1987–1988. Washington, DC. pp. 1–14.

Formisamo, Ronald P./Burns, Constance K. (1984): Boston 1700–1980. The Evolution of Urban Politics. Westport.

Foundation for Architecture (ed.) (1991): Philadelphia Architectural Timeline. Philadelphia.

Frank, Karolin (1997): Philadelphia – postindustrieller Phönix aus der Asche. Revitalisierung, Aufwertung und Denkmalpflege. Staatsexamensarbeit, Universität Hamburg.

Frank, Karolin (1998): Philadelphia – Aufwertung, Revitalisierung, Denkmalpflege. In: Arbeitskreis der Deutschen Gesellschaft für Geographie: USA. Mitteilungsblatt No. 24/1998. Lüneburg. pp. 24–48.

Frank, Karolin (1999): Erforschung der Denkmalpflege in den USA. Probleme und Potentiale im urbanen Raum. In: Mitteilungsblatt des Arbeitskreis der Deutschen Gesellschaft für Geographie: USA. Mitteilungsblatt No. 26/1999. Lüneburg. pp. 15–25.

Frank, Karolin (2001): Die Umnutzung von Gebäuden und Industriebrachen in den USA. Stadtplanerische und gesellschaftliche Aspekte. In: Mitteilungen der Geographischen Gesellschaft Hamburg. F. N. Nagel (ed.). Band 91. Hamburg.

Frantz, John B./Pencak, William (1998): Beyond Philadelphia. The American Revolution in the Pennsylvania Hinterland. Pennsylvania State University Press. University Park.

Friedrichs, Jürgen (1995a): Methoden empirischer Sozialforschung. Verlag Leske + Budrich. Opladen.

Friedrichs, Jürgen (1995b): Stadtsoziologie. Verlag Leske + Budrich. Opladen.

Friedrichs, Jürgen (1997): Die Individualisierungsthese. Verlag Leske + Budrich. Opladen.

Friedrichs, Jürgen/Kecskes, Robert (1996): Gentrification: Theorie und Forschungsergebnisse. Verlag Leske + Budrich. Opladen.

Friend, John/Hickling, Allen (eds.) (^{2}1997): Planning Under Pressure: The Strategic Choice Approach. Oxford, UK.

Fuqua, Jonathan Scott (1997): American Rowhouses Classic Designs. Owings Mill, Maryland.

Gallery, John Andrew (^{2}1994): Philadelphia Architecture: A Guide to the City. Philadelphia.

Gantt, Harvey (1993): Reassessing Our Agenda (Focus on Cultural Diversity II). In: National Trust for Historic Preservation (ed.): Historic Preservation Forum. No. 1–2/1993. Washington, DC. pp. 6–11.

Gamerith, Werner (1999): Hochqualifizierte Personen aus ethnischen Minderheitengruppen in den USA. Räumliche Divergenzen und soziale Konfliktfelder. In: Arbeitskreis der Deutschen Gesellschaft für Geographie: USA. Mitteilungsblatt No. 25/1999. Lüneburg. pp. 20–32.

Garvin, Alexander (1996): The American City: What Works, What Doesn't. New York.

Gelfand, Mark I. (1990): Boston. Back to the Politics of the Future. In: Bernhard, Richard M. (ed.): Snowbelt Cities. Metropolitan Politics in the Northeast and Midwest. University of Indiana Press. Bloomington. pp. 40–62.

Genius-Devime, Barbara (1996): Bedeutung und Grenzen des Erbes der Menschen im völkerrechtlichen Kulturgüterschutz. In: Saarbrücker Studien zum Internationalen Recht. Nomos-Verlag. Baden-Baden.

Goldfield, David R. (1990): The Stages of American Urbanization. In: Magazine of History. No. 3/1990.

Goldfield, David R./Brownell, Blaine A. (eds.) (1979): Urban America: From Downtown to No Town. Boston.

Goode, Judith/Schneider, Jo-Anne (1994): Reshaping Ethnic and Racial Relations in Philadelphia. Temple University Press. Philadelphia.

Göschel, Albrecht/Kirchberg, Volker (eds.) (1998): Kultur in der Stadt. Stadtsoziologische Analysen zur Kultur. Verlag Leske + Budrich. Opladen.

Gratz, Roberta Brandes (1994): The Living City. How American Cities are Being Revitalized by Thinking Small in a Big Way. Preservation Press. Washington, DC.

Greater Philadelphia Chamber of Commerce (1993): Greater Philadelphia Get to Know Us. No. 3/1993. Philadelphia.

Gross, Sandra/Roberts, Jeffrey P. (1983): William Penn: Visionary & Proprietor. Philadelphia.

Gutek, Gerald Lee (1994): Experiencing America's Past: A Travel Guide to Museum Villages. University of South Carolina Press. Columbia.

Gutschow, Niels (1978): Denkmalpflege – Stadterhaltung in den USA. In: Deutsche Kunst- und Denkmalpflege. Jg. 36, 1978. München, Berlin. pp. 131–142.

Gough, Debrah Mathias (1995): Christ Church, Philadelphia. University of Pennsylvania Press. Philadelphia.

HABS – Historic American Buildings Survey (ed.) (1983): Historian's Procedures Manual. Washington, DC.

HABS/HAER – Historic American Buildings Survey/Historic American Engineering Records (1983): Historic America. Buildings, Structures, and Sites. Washington, DC.

Hagget, Peter (1990): The Geographer's Art. Blackwell Publisher. Cambridge, Massachusetts.

Hahn, Roland (1990): USA. Klett Länderprofile. Stuttgart.

Hall, Peter Geoffrey (1966): The World Cities. New York.

Hall, Peter Geoffrey (^{6}1997): Cities of Tomorrow: An Intellectual History of Urban Planning and Design in the Twentieth Century. Basil Blackwell Ltd. Oxford.

Halverson-Pace, Valerie Sue (1976): Society Hill Philadelphia: Historic Preservation and Urban Renewal in Washington Square East (Volume I and II). Ph.D. Dissertation. University of Minnesota.

Hammer, Felix (1995): Die geschichtliche Entwicklung des Denkmalrechts in Deutschland. In: Heckel, Martin et al. (eds.): Jus Ecclesiasticum. Beiträge zum evangelischen Kirchenrecht und zum Staatsrecht. Band 51. Tübingen.

Harris, Richard/Lewis, Robert (1998): How the Past Matters: North American Cities in the Twentieth Century. In: Journal of Urban Affairs. Vol. 20, No. 2/1998, pp. 159–174.

Harth, Annette et al. (1986): Ausgewählte Segregationsprozesse in ostdeutschen Großstädten, dargestellt am Beispiel der Landeshauptstadt Magdeburg. In: Stadtforschung und Statistik. No. 2/1999. pp. 20–29.

Hayden, Dolores (^{6}1986): Redesigning the American Dream: The Future of Housing, Work and Family Life. New York.

Hayes, Bob (1973): The Black American Travel Guide. San Francisco.

Historic Buildings & Monuments Commission for England (1996): English Heritage Guide. Tunbridge, UK.

HBI – Historic Boston Incorporated (ed.) (1988): A Case Study on the Meditation Process Over the Preservation of a Religious Building: The Church of the Immaculate Conception: Boston. Boston.

HBI – Historic Boston Incorporated (ed.) (1991): Religious Property Preservation: A Boston Casebook. Boston.

Heinritz, Günther/Helbrecht, Ilse (eds.) (1998): Sozialgeographie und Soziologie. Dialog der Disziplinen. Münchner Geographische Hefte Band 78. LIS-Verlag. Passau.

Heinz, Kersten (1991): Kultur, Kulturbegriff, Kulturdenkmalbegriff. Peter Lang Verlag. Frankfurt am Main.
Herbert, David T. (1995): Heritage, Tourism and Society. London, New York.
Heritage Foundation (ed.) (1985): A New Agenda for Education. Washington, DC.
Historical Atlas of Religion in America (1962): Gaustad, Edwin Scott (ed.). New York.
Historical Publication Society in Philadelphia (1926): History of Philadelphia. Philadelphia.
Historic Preservation. A Guide to Information Sources (o.J). Markowitz, Arnold L. (ed.). Detroit.
Hodder, Robert (1999): Redefining a Southern's City's Heritage. In: Journal of Urban Affairs. Vol. 21, No. 4/1999. pp. 437–453.
Hodler, T. W. (1986): The Atlas of Georgia. University of Georgia Press. Athens, Georgia.
Hofmeister, Burkhard (1970): Nordamerika. Fischer Länderkunde. Band 6. Frankfurt am Main.
Hofmeister, Burkhard (1977): Die Erhaltung historisch wertvoller Bausubstanz in den Städten der USA. In: Die Erde. No. 108/1977. pp. 129–150.
Hofmeister, Burkhard (1982): Die Stadtstruktur im interkulturellem Vergleich. In: Geographische Rundschau No. 34/1982, Heft 11. pp. 482–488.
Hofmeister, Burkhard (1986): Denkmalpflege und Denkmalschutz in Australien. In: Die Alte Stadt. No. 2/1986. pp. 81–98.
Hofmeister, Burkhard (1989): Stadtgeographie. Entwicklungsphasen und wechselnde Forschungsschwerpunkte. In: Die Alte Stadt. No. 2–3/1989. pp. 411–420.
Hofmeister, Burkhard (^{6}1993): Stadtgeographie. Braunschweig.
Hofmeister, Burkhard (^{7}1999): Stadtgeographie. Braunschweig.
Holleran, Michael (1998): Boston's Changeful Times. Origins of Preservation & Planning in America. John Hopkins University Press. Baltimore.
Hoppenfeld, Morton (1960): The Role of Design in City Planning – With Reference to Center-City Philadelphia. In: Journal of the American Institute of Planners. Vol. 26, No. 2/1960. pp. 98–103.
Hosmer, Charles Bridgham (1965): Presence of the Past: A History of the Preservation Movement in the United States Before Williamsburg. New York.
Hosmer, Charles Bridgham Jr. (1981): Preservation Comes of Ages: From Williamsburg to the National Trust. 1926–1949. University of Virginia Press. Charlottesville.
HSF – Historic Savannah Foundation (1965): Savannah as a Travel Destination. Savannah.
HSF – Historic Savannah Foundation (^{2}1979): Survey of Significant Buildings in the National Historic District. Savannah.
Huber, Gitte (1996): Denkmalpflege zwischen Kunst und Wissenschaft. Arbeitshefte des Bayerischen Landesamtes für Denkmalpflege. Band 76. München.
Hunter, Ann Arnold (1991): A Century of Service: The Story of DAR. Washington, DC.
Hutter, Michael (1997): Economic Perspectives on Cultural Heritage. New York.
ID – Informationsdienst und Mitteilungsblatt des VHW-Deutsches Volksheimstättenwerk e.V. (ed.) (1999): Hinter die Fassaden gucken: Der Wohnungs- und Städtebau im Zukunftsprogramm 2000 der Bundesregierung. 53. No. 15/1999. pp. 149–163.
ILS – Institut für Landes- und Stadtentwicklungsforschung des Landes Nordrhein-Westfalen (ed.) (1987): Umnutzung von Baudenkmälern. Eine Planungshilfe für die Praxis. Bausteine 19. Dortmund.
ILS – Institut für Landes- und Stadtentwicklungsforschung des Landes Nordrhein-Westfalen (ed.) (1988): Baudenkmäler in Nordrhein-Westfalen. ILS 14. Dortmund.

ILS – Institut für Landes- und Stadtentwicklungsforschung des Landes Nordrhein-Westfalen (ed.) (1994): Neues Wohnen in alten Mauern. Beispiele für die Umnutzung wertvoller Bausubstanz. Bausteine 9. Dortmund.

ILS-Institut für Landes- und Stadtentwicklungsforschung des Landes Nordrhein-Westfalen (ed.) (1999): Integrierte Stadtteilerneuerung und Bewohneraktivierung in nordamerikanischen Großstädten. ILS 142. Dortmund.

Informationen zur politischen Bildung (ed.) (1997) : Das politische System der USA. Bundeszentrale für politische Bildung. Bonn.

Jackson, Kenneth T. (1985): Crabgrass Frontier. The Suburbanization of the United States. Oxford University Press. New York.

Jacobs, Jane (1984): Cities and the Wealth of Nations: Principles of Economic Life. Random House. New York.

Jacobs, Jane (^{3}1993): The Death and Life of Great American Cities. New York.

Jacobs, Stephen W. (1966): Architectural Preservation in Europe: French and English Contributions. Cornell University. Discussion paper No. 25. Ithaca.

Jaher, Frederic Cople (1982): The Urban Establishment. University of Illinois Press. Urbana.

Jodidio, Philip (1997): Contemporary American Architects. Volume III. Taschen Verlag. Köln.

Juliani, Richard N. (^{2}1999): Building Little Italy. Philadelphia's Italians Before the Mass Migration. Penn State University Press. Harrisburg.

Johnston, R. J. (1978): Residential Area Characteristics Research Methods for Identifying Urban Subareas. In: Social Era Analysis. Factorial Ecology. London.

Kane, Joseph Nathan/Alexander, Gerard L. (^{3}1979): Nicknames and Soubriquets of U.S. Cities, States, and Counties. Metuchen, New Jersey.

Kapp, William K. (1950): The Social Costs of Private Enterprise. Cambridge, Massachusetts.

Karutz, Michael (1998): Renaissance der Innenstädte durch Integration und Vernetzung? In: Die PlanerIn. No. 3/1998. pp. 24–25.

Katz, Peter (1994): The New Urbanism: Toward an Architecture of Community. Portland.

Katz, Michael (1998): Redefining the Welfare State in America, 1980–1999. Lecture manuscript, University of Pennsylvania. Philadelphia.

Kaufmann, Arthur C. (1967): The Economic Feasibility of the Chestnut Street Mall in Philadelphia. In: Praeger, Frederick A. (ed.): Taming Megalopolis. Vol. 2. New York.

Kashatus, William C. (1998): Preserving Philadelphia: Neighborhoods, Division and Conflict in a Postindustrial City. A Conversation With Charles E. Peterson F.A.I.A. In: Heritage Pennsylvania. Vol. 26, No. 4/1998. Harrisburg. pp. 22–31.

Kegler, Harald (1998): Mehr Sehnsucht nach der alten Stadt: New Urbanism in den USA. In: Die Alte Stadt. No. 4/1998. pp. 355–346.

Kelsey, Elisabeth Allison (1997): How the Other Nine-Tenths Lived: Interpreting the Working Class Experience in Philadelphia 1870–1900. Unpublished M.A thesis. University of Pennsylvania. Philadelphia.

Kerstenbaum, Herbert (ed.) (1994): Regional Almanac 1995. Philadelphia.

Kiesow, Gottfried (^{3}1995): Einführung in die Denkmalpflege. Darmstadt.

Kiesow, Gottfried (1999): Kulturgeschichte sehen lernen. Deutsche Stiftung Denkmalschutz. Monumente Kommunikation. Bonn.

Kirchberg, Volker/Göschel, Albrecht (eds.) (1998): Kultur in der Stadt. Stadtsoziologische Analysen zur Kultur. Verlag Leske + Budrich. Opladen.

Kolakowsky, Lisa Marie (1997): Historic Structures in the Redevelopment Process Market Street East, Philadelphia. Unpublished M.A thesis. University of Pennsylvania. Philadelphia.

Korff, Rüdiger (1991): Die Weltstadt zwischen globaler Gesellschaft und Lokalitäten. In: Zeitschrift für Soziologie. Jg. 20, Heft 5/1991. pp. 357–368.

Korn, Karl (1978): Nur Denkmalschutz? In: Deutsche Kunst- und Denkmalpflege. Jg. 36, 1978. München, Berlin. pp. 143.

Korte, Hermann (1989): Statik und Prozeß. Rede vom Wesen der Städte. In: Die Alte Stadt. No. 2–3/1989. pp. 173–185.

Koshar, Rudi (1998): Germany's Transit Pasts: Preservation and National Memory in Twentieth Century. University of North Carolina Press. Durham.

Krämer, Walter (1997): Statistische Probleme bei der Armutsmessung. Gutachten im Auftrag des Bundesministeriums für Gesundheit. In: Schriftenreihe des Bundesministeriums für Gesundheit. Band 94. Nomos Verlagsgesellschaft. Baden Baden.

Krier, Bob (^{5}1991): Urban Space. New York.

Krueckeberg, Donald A. (^{3}1997): Introduction to Planning History in the United States. New Brunswick.

Kunstler, James Howard (1994): The Geography of Nowhere: The Rise and Decline of America's Man – Made Landscape. New York.

Kunstler, James Howard (1998): Home from Nowhere: Remaking our Everyday World for the 21st Century. New York.

Kunze, Ronald/Kurth, Detlef (1998): Stadt von innen – statt nach außen. Strategien der Stadterneuerung für die Zukunft der Stadt. In: Die PlanerIn. 2/98. pp. 3.

Kupinski, Hans-Dieter (1999): Wohnen in alten Mauern. Umnutzung ehemaliger Nichtwohngebäude in Wohnraum. In: vhw-Informationsdienst. No. 13/1999. pp. 136–142.

Landesamt für Denkmalpflege Hessen (1996): Denkmalpflege in Hessen. Wiesbaden.

Langdon, Philip (1994): A Better Place to Live: Reshaping the American Suburb. University of Massachusetts Press. Amherst, Massachusetts.

Larkham, Peter J./Jones, Andrew N. (eds.) (1991): A Glossary of Urban Form. In: Historical Geography Research Series. No. 26. Washington, DC.

Laux, Hans Dieter/Thieme, Günther (1998): Zwischen Immigration und Integration: Asiatische Einwanderer in den USA. In: DFG – Informationen. No. 4/1998. Bonn. pp. 12–14.

Leach, Sara Amy (ed.) (1989): Two Historic Pennsylvania Canal Tours: Alexandria and Saltsburg. Washington, DC.

Leopold, Ellen (1991): Boston's Central Artery/Third Harbor Tunnel Project. In: Urban Design and Preservation Quarterly. Vol. 14, No. 2/1991. pp. 1–5.

Lerner, Shereen (1994): Introduction: Historic Preservation and Archeology. In: National Trust for Historic Preservation (ed.): Historic Preservation Forum. No. 3–4/1994. Washington, DC. pp. 4–6.

Levine, Hillel (1992): The Death of an American Jewish Community: A Tragedy of Good Intentions. New York.

Levine, Marc V. (1987): Downtown Redevelopment as an Urban Growth Strategy: A Critical Appraisal of Baltimore Renaissance. In: Journal of Urban Affairs. Vol. 9, No. 2/1987. pp. 103–123.

Lichtenberger, Elisabeth (1986): Stadtgeographie – Perspektiven. In: Geographische Rundschau. No. 38/1986, Heft 7/8. pp. 388–394.

Lichtenberger, Elisabeth (1989): Stadtentwicklung in Europa und Nordamerika – kritische Anmerkungen zur Konvergenztheorie. In: Heyer, R./Hommel, M. (eds.): Stadt und Kulturraum. Bochumer Geographische Arbeiten. Nummer 50. Bochum. pp. 113–129.

Lichtenberger, Elisabeth (1990): Stadtverfall und Stadterneuerung. In: Beiträge zur Stadt- und Regionalforschung. Band 10. Wien.

Lichtenberger, Elisabeth (1992): Die Großstadt in der postindustriellen Gesellschaft. In: Die Stadt als Kultur- und Lebensraum. Vorträge im Wintersemester 1990/1. Heidelberg. pp. 9–24.

Lichtenberger, Elisabeth (^{3}1998): Stadtgeographie I. Begriffe, Konzepte, Modelle, Prozesse. Teubner Studienbücher. Stuttgart, Leipzig.

Lipset, Seymour Martin (1981): The Third Century. America as a Post-Industrial Society. Hoover Institution Press. Stanford University. Stanford.

Lindgren, James Michael (1995): Preserving Historic New England: Preservation, Progressivism and the Remaking of Memory. Oxford University Press. London, New York.

Longstreth, Richard W./Teitelmann, Edward (^{2}1981): Architecture in Philadelphia: A Guide. Cambridge, Massachusetts.

Looney, Robert, F. (1976): Old Philadelphia in Early Photographs 1839–1914. New York.

Looney, Robert, F. (ed.) (1983): Picture Postcards of Old Philadelphia. Mineola, New York.

Lorenz, Gert (1999): Denkmalgeschützte Immobilien: Zusammenarbeit ist gefragt. In: Immobilienmanager. No. 7–8/1999. pp. 12–24.

Lowe, David (ed.) (^{2}1982): Old Boston. Mineola, New York.

Lowenthal, David (^{3}1988): The Past is a Foreign Country. Cambridge University Press. New York.

Lowenthal, David/Bowden, Martyn J. (eds.) (1976): Geographies of the Mind. Oxford University Press. London, New York.

Luckmann, Thomas (1996): The Privatization of Religion and Morality. In: Heelas, Paul et. al. (eds.): Historic Preservation – Traditionalization: Critical Reflections on Authority and Identity at a Time of Uncertainty. Blackwell Publisher. Cambridge, Massachusetts.

Lynch, Kevin (^{2}1982): A Theory of Good City Form. MIT Press. Cambridge, Massachusetts.

Lynch, Kevin (^{25}1997): The Image of the City. MIT Press. Cambridge, Massachusetts.

Marcuse, Peter (1986): Abandonment, Gentrification and Displacement: The Linkage in New York City. In: Smith, Neil/Williams, Peter (eds.): Gentrification of the City. Winchester.

Marcuse, Peter (1998): Ethnische Enklaven und rassische Ghettos in der postfordistischen Stadt. In: Heitmeyer, Wilhelm et al. (eds.): Die Krise der Städte. Analysen zu den Folgen desintegrativer Stadtentwicklung für das ethnisch-kulturelle Zusammenleben. Frankfurt am Main.

Martin Greenwald Associates (ed.) (1984): Historical Maps on File. Facts on File. New York.

Marquis, Kelly Deborah/Goodman, Jennifer (1993): Conservation Districts as Alternatives to Historic Districts. In: National Trust for Historic Preservation (ed.): Historic Preservation Forum. No. 9–10/1993. Washington, DC. pp. 6–14.

Matthews, Melinda J. (1992): Affordable Housing and Historic Preservation. In: National Trust for Historic Preservation (ed.): Historic Preservation Forum. No. 5–6/1992. Washington, DC. pp. 6–11.

Maxmann, Susan A. (1994): Historic Preservation: A Sustainable Strategy for the ´90s. In: National Trust for Historic Preservation (ed.): Historic Preservation Forum. No. 1–2/1994. Washington, DC. pp. 4–6.

McCall, Elisabeth B. (1966): Old Philadelphia Houses on Society Hill 1750–1840. New York.

Menino, Thomas N. (1995): Boston and the Main Street Program. Using the Rosalinde Village Approach to Revitalize Neighborhood Commercial Center City – Wide. In: National Trust for Historic Preservation (ed.): Historic Preservation Forum. No. 2/1995 . Washington, DC. pp. 46–48.

Mercer, James L. (1991): Strategic Planning for Public Managers. Quorum Books. New York, London.

Millard, Kent R. (1991): Fast – Food Chains: A Review of Their Evolution and Visual Impact on Communities. In: National Trust for Historic Preservation (ed.): Historic Preservation Forum. No. 1–2/1991. Washington, DC. pp. 36–47.

Miller, E. Willard (1962): A Geography of Manufacturing. Englewood Cliffs, New Jersey.

Miller, Page Putnam (ed.) (1992): Reclaiming the Past: Landmarks of Women's History. Washington, DC.

Moak, Jefferson M. (1996): Philadelphia Street Names Change. Philadelphia.

Mollenkopf, John Hull (1992): A Phoenix in the Ashes: The Rise and Fall of the Koch Coalition in New York Cities Politics. Princeton.

Möller, Hans-Herbert (1983): Niedersächsische Denkmalpflege: Berichte über die Tätigkeit der Bau- und Kunstdenkmalpflege in den Jahren 1979–1982. Hannover.

Morley, Christopher (^{2}1992): Philadelphia. Fordham, New York.

Morgan, George (1926): History of Philadelphia. Philadelphia.

Mooney-Melvin, Patricia (1996): Urban History, Local History, Public History. In: History News. Vol. 51, No. 2/1996. Nashville.

Moos, Roger W. (1998): Historic Houses of Philadelphia. University of Pennsylvania Press. Philadelphia.

Murtagh, William J. (1997): Keeping Time: The History and Theory of Preservation in America. New York.

Murphy, Raymond E. (1971): The Central Business District. London, UK.

Murphy, Raymond E./Vance Jr., J. E. (1954): Delimiting the C.B.D. In: Economic Geography 1954. pp. 189–222.

Nagel, Frank Norbert (ed.) (1995): Stadtentwicklung und Stadterneuerung. Mitteilungen der Geographischen Gesellschaft in Hamburg. Band 85. Stuttgart.

Nagel, Frank Norbert (1999): Kulturlandschaftsforschung und Industriearchäologie. Inhalte, Methodik, Anwendung. In: Budesheim, Werner et al. (eds.): Zur Kulturgeographie und Industriearchäologie in Norddeutschland. Versorgung, Verteidigung, Verkehr. Beiträge zur Wissenschaft und Kultur. Band 4. Wentorf bei Hamburg. pp. 9–25.

Nagel, Frank Norbert/Goldammer, Götz (1997): Wasserwege als Gegenstand der Kulturlandschaftspflege. In: Schenk, Winfried et al. (eds.): Kulturlandschaftspflege. Beiträge der Geographie zu räumlichen Planung. Berlin. pp. 275–285.

Nagel, Frank Norbert/Oberbeck, Gerhard (1982): Neue Formen städtischer Entwicklung im Südwesten der USA: Sonnenstädte der zweiten Generation. In: Mitteilungen der Geographischen Gesellschaft Hamburg. Band 72. Hamburg.

Nahr, Heinrich (1999): Das Programm der ‚Sozialen Stadtteilentwicklung'. In: Hamburger Sozialhilfe Report. No. 16/1999. pp. 18–21.

National Park Service (ed.) (1998): The Black Heritage Trail. Boston.

National Park Service/Department of the Interior (eds.) (1998): African-American National Historic Site. Boston.

National Park Service/Department of the Interior/American Folklife Center/Library of Congress (eds.) (1983): Cultural Conservation: The Protection of Cultural Heritage in the United States: A Cooperative Study. Washington, DC.

National Park Service/National Trust for Historic Preservation (eds.) (1994): National Register of Historic Places 1966 to 1994. Preservation Press. Washington, DC.
New York Times (ed.) (1976): The Cities. Arno Press. New York.
NN (n. d.): Historic Savannah, Georgia. Savannah.
NN (^{3}1998): Access. Philadelphia. Makes the World Your Neighborhood. New York.
Norcliffe, Glen et al. (1996): The Emergence of Postmodernism on the Urban Waterfront. In: Journal of Transport Geography. Vol. 4, No. 2/1996. pp. 123–134.
NSDAR – National Society of the Daughters of the American Revolution (1979): Historic and Memorial Buildings of the Daughters of the American Revolution. Washington, DC.
NSDAR – National Society of the Daughters of the American Revolution (1996): Handbook and National Bylaw. Washington, DC.
NTHP – National Trust for Historic Preservation (1975): Economic Benefits of Preserving Old Buildings. Papers from the Economic Benefits of Preserving Old Buildings Conference in Seattle, Washington (State).
NTHP – National Trust for Historic Preservation (1980): Preservation: Toward an Ethic in the 1980s. Preservation Press. Washington, DC.
NTHP – National Trust for Historic Preservation (ed.) (^{3}1983): With Heritage so Rich. New York.
NTHP – National Trust for Historic Preservation (1987): The Architectural Survey. Washington, DC.
NTHP – National Trust for Historic Preservation (ed.) (1995): The Promise of Heritage Tourism. In: Historic Preservation Forum News. No. 1–2/1995. Washington, DC.
NTHP – National Trust for Historic Preservation (ed.) (1996a): Historic Preservation Forum Index Volumes 1–10, 1987–1995. In: Historic Preservation Forum. Washington, DC.
NTHP – National Trust for Historic Preservation (ed.) (1996b): Information About the National Trust for Historic Preservation. Washington, DC.
NTHP – National Trust for Historic Preservation (1997): Preservation Yellow Pages: The Complete Information Source for Homeowners, Communities and Professionals. Washington, DC.
NTHP – National Trust for Historic Preservation (ed.) (1998a): Authentic Heritage Tourism. In: Historic Preservation Forum News. No. 3–4/1998. Washington, DC.
NTHP – National Trust for Historic Preservation (ed.) (1998b): Economic Benefits of Historic Preservation . In: Preservation Forum News. No. 5–6/1998. Washington, DC.
NTHP – National Trust for Historic Preservation (ed.) (1998c): Six Cities: Pioneers in Preservation. In: Preservation Forum News. No. 4/1998. Washington, DC.
OEDC – Organisation for Economic Co-Operation and Development (ed.) (1998): Integrating Distressed Urban Areas. Paris.
Ott, Thomas (1999): Vortrag Deutscher Geographentag in Hamburg. www.geographie.uni-mannheim.de/ott/ vanseagt99.html (information from 8th April, 2000).
Palen, John J. (1975): The Urban World. New York.
Partners for Sacred Places (ed.) (1998): Sacred Places at Risk. Philadelphia.
PCVB – Philadelphia Convention & Visitors Bureau (1997): Philadelphia Business & Leisure Travel Report. Philadelphia.
PEL – Pennsylvania Economy League (ed.) (1989): Prospects for Linking Commercial and Community Development in Philadelphia. Philadelphia.
Pennsylvania Historical and Museum Commission (ed.) (1988). WPA-Guide to Philadelphia. University of Pennsylvania Press. Philadelphia.

Petzet, Michael/Mader, Gert (1993): Praktische Denkmalpflege. Kohlhammer Verlag. Stuttgart.
PHC – Philadelphia Historical Commission (ed.) ([2]1997): Rules and Regulations. Philadelphia.
PHC – Philadelphia Historical Commission (ed.) (n. d.): Philadelphia Historic Districts. Philadelphia, Pennsylvania.
Phil Plan – Philadelphia City Planning Commission (ed.) (various volumes): Annual Report. Philadelphia.
Phil Plan – Philadelphia City Planning Commission (ed.) (1960a): The Physical Development Plan for the City of Philadelphia 1960. Philadelphia.
Phil Plan – Philadelphia City Planning Commission (ed.) (1960b): City of Philadelphia: Existing Land Use. Philadelphia.
Phil Plan – Philadelphia City Planning Commission (ed.) (1963): Center City Redevelopment Area Plan. Philadelphia.
Phil Plan – Philadelphia City Planning Commission (ed.) (1974): Center City Redevelopment Area Plan. Philadelphia.
Phil Plan – Philadelphia City Planning Commission (ed.) (1982a): Central Riverfront District Plan. Philadelphia.
Phil Plan – Philadelphia City Planning Commission (ed.) (1982b): Economic and Social Indicators for Philadelphia Census Tracts – 1980. Philadelphia.
Phil Plan – Philadelphia City Planning Commission (ed.) (1983): Population and Housing Characteristics for Philadelphia Census Tracts – 1980 & 1970. Philadelphia.
Phil Plan – Philadelphia City Planning Commission (ed.) (1984a): Socio-Economic Characteristics for Philadelphia Census Tracts – 1980 & 1970. Philadelphia.
Phil Plan – Philadelphia City Planning Commission (ed.) (1984b): Center City Housing Market Study: Conclusion. Philadelphia.
Phil Plan – Philadelphia City Planning Commission (ed.) (1984c): Center City Housing Market Study: Background/Summary. Philadelphia.
Phil Plan – Philadelphia City Planning Commission (ed.) (1986): The Future of Center City–Three Scenarios. Philadelphia.
Phil Plan – Philadelphia City Planning Commission (ed.) (1987): The Plan for Center City: Historic Preservation. Philadelphia.
Phil Plan – Philadelphia City Planning Commission (ed.) (1988a): The Plan for Center City. Philadelphia.
Phil Plan – Philadelphia City Planning Commission (ed.) (1988b): The Urban Form of Center City. Philadelphia.
Phil Plan – Philadelphia City Planning Commission (ed.) (1990a): Industrial Philadelphia: A Study of Land Use. Philadelphia.
Phil Plan – Philadelphia City Planning Commission (ed.) (1990b): New C-4 and C-5 Commercial Districts. Philadelphia.
Phil Plan – Philadelphia City Planning Commission (ed.) (1990c): Investing in Philadelphia. The 1991–1996 Capital Program. Philadelphia.
Phil Plan – Philadelphia City Planning Commission (ed.) (1990d): Market Street East Urban Design Study. Philadelphia.
Phil Plan – Philadelphia City Planning Commission (ed.) (1990e): The Plan for Center City: Report on Implementation. Philadelphia.
Phil Plan – Philadelphia City Planning Commission (ed.) (1991a): Preliminary Population Data for Census Tracts. Philadelphia.

Phil Plan – Philadelphia City Planning Commission (ed.) (1991b): Selected Tables: Population and Housing Data: 1990&1980. Philadelphia.
Phil Plan – Philadelphia City Planning Commission (ed.) (1991c): List of Publications. Philadelphia.
Phil Plan – Philadelphia City Planning Commission (ed.) (1992). Zoning Maps: Building and Zoning Version. No. 46, 47, 60, 61. Philadelphia.
Philadelphia Housing Association (ed.) (1961): Ends and Means of Urban Renewal. Philadelphia.
PIDC – Philadelphia Industrial Development Corporation (1993): Philadelphia. Philadelphia.
Poston, Johathan H. (1997): The Buildings of Charleston. A Guide to the City. University of South Carolina Press. Columbia.
Potter, Parker B. (1994): Public Archeology in Annapolis: A Critical Approach to History in Maryland's Ancient City. Smithsonian Institution Press. Washington, DC.
Purdy, Lisa A. (1990): The Anatomy of Creating a Local Historic District. In: Urban Design and Preservation Quarterly. Vol. 13, No. 2/1990. pp. 8–11.
Praeger, Frederick A. (ed.) (1967): Taming Megalopolis. Band 2. New York.
Precht von Taborizki, Barbara (1996): Die Denkmallandschaften. Ensemble, schützenswerte Gesamtheit, Denkmalumgebung. Köln.
Prentice, Richard (1993): Tourism and Heritage Attractions. Routledge Publisher. New York.
Preservation Alliance for Greater Philadelphia (ed.) (1998a): The Rittenhouse Fitler Historic District Manual. A Guide for Property Owners. Philadelphia.
Preservation Alliance for Greater Philadelphia (ed.) (1998b): The Economic Benefits of Preserving Philadelphia's Past. Philadelphia.
Preservation Alliance for Greater Philadelphia (ed.) (n. d.): How to Prepare a Nomination to the Philadelphia Register of Historic Places. Philadelphia.
Preston, Andrew (1990): Philadelphia. A Picture Memory. Singapore.
Rath, Frederick L./O´Connel, Merrylyn Rogers (ed.) (1975): Historic Preservation. A Bibliography on Historical Organization Practices. Nashville.
RDA – Redevelopment Authority Philadelphia (1968): Annual Report 1968. City of Philadelphia (ed.). Philadelphia.
RDA – Redevelopment Authority Philadelphia (1969): Annual Report 1969. City of Philadelphia (ed.). Philadelphia.
RDA – Redevelopment Authority Philadelphia (1980): Annual Report 1980. City of Philadelphia (ed.). Philadelphia.
RDA – Redevelopment Authority Philadelphia (1981): Annual Report 1981. City of Philadelphia (ed.). Philadelphia.
Reul, Horst (1998): Bauliches Erbe in der Stadt. Die ideell – symbolische und ökonomische Dimension lokaler Denkmalpolitik. Dissertation. Universität Augsburg.
Rhoad, Debbi (1994): The Board of Architectural Review in Charleston, 1931–1993. Charleston.
Ripley, Warren (1996): Charles Towne: Birth of a City. Charleston.
Rizza, Paul F./Hughes, James C. (^{2}1982): Pennsylvania Atlas. A Thematic Atlas of the Keystone State. Grove City.
Robinson, Kathy (1983): Cities of the World: World Regional Urban Development. New York.
Röller, Dirk (1996): Stadt und Mensch zwischen Chaos und Ordnung: Referate des Internationalen Semiotischen Symposiums zur Stadtanthropologie 1994. Frankfurt am Main.

Roth, Stacy (1998): Past Into Present: Effective Techniques for First – Person Historical Interpretation. University of North Carolina Press. Chapel Hill.

Row, Arthur (1960): The Physical Development Plan. In: Journal of the American Institute of Planners. Vol. 26, No. 3/1960. pp. 177–185.

Rucht, Dieter (1994): Modernisierung und heutige soziale Bewegungen. Frankfurt am Main.

Samuels, Gayle B. (^{2}1996): Women in the City of Brotherly Love .. and Beyond. Philadelphia.

Sauder, Robert A./Wilkinson, Teresa (1989): Preservation Planning and Geographic Change in New Orleans' Vieux Carré. In: Urban Geography. No. 10/1989. pp. 41–61.

Sassen, Saskia (1991): The Global City: New York, London, Tokyo. Princeton.

Sassen, Saskia (1994): Cities in a World Economy. Thousand Oaks.

Savannah Housing Authority (1964): Historic Renewal Program for Savannah. Savannah.

Savannah Visitors Bureau (1999): Savannah Historic District. Map and Legend. Savannah.

Sautter, Udo (^{3}1986): Geschichte der Vereinigten Staaten von Amerika. Stuttgart.

Schätzel, Ludwig (^{2}1994): Wirtschaftsgeographie Band 2. Paderborn.

Schneider, Adam David (1990): Practical Rehabilitation: A Guide for the Revitalization of Older, Urban Neighborhoods. Unpublished Master's thesis. University of Pennsylvania. Philadelphia.

Schneider-Sliwa, Rita (1995): Stadtentwicklung für das 21. Jahrhundert – das Modell Boston. In: Geographische Rundschau Jg. 47, Heft 10/1995. pp. 586–593.

Schubert (1977): Soziale Stadtplanung in Wohnquartieren. Frankfurt am Main.

Schuck-Wersig, Petra/Wersig, Gernot (1999): Museumsmarketing in den USA. Neue Tendenzen und Erscheinungsformen. In: Berliner Schriften zur Museumskunde Band 15. Berlin.

Schuster, Mark J. et al. (eds.) (1997): Preserving the Built Heritage: Tools for Implementation. Salzburger Seminar. University Press of New England. Hanover, New Hampshire.

Schwartz, Alvin (1968): Old Cities & New Towns. Changing Face of the Nation. New York.

Scully, Vincent (^{2}1988): American Architecture and Urbanism. New York.

Selhat, Lynn (1995): Penn's Landing. Philadelphia's Birthplace Reborn. In: Urban Land Institute. No. 319. Washington, DC. pp. 47–50.

Shaw, Brian J./Jones, Roy (1997): Contested Urban Heritage: Voices From the Periphery. Hants.

SHCA – Society Hill Civic Association (1994): Preservation Committee: Chilton's 1994 Guide to Historic Society Hill. Philadelphia.

SHCA – Society Hill Civic Association (1999): The Society Hill Reporter. No. 5/1999. Philadelphia.

Siegel, Adrienne (1975): Philadelphia. Philadelphia.

Siegel, Michael (1984): Denkmalpflege als öffentliche Aufgabe. In: Beiträge zur Stadt- und Regionalforschung Band 13. Hamburg.

Sieverts, Thomas (1998): Die Stadt in der zweiten Moderne, eine europäische Perspektive. Informationen zur Raumentwicklung. Heft 7–8/1998. Bonn.

Silver, Christopher (1991): Revitalizing the Urban South. Neighborhood Preservation and Planning Since the 1920s. In: Journal of the American Planning Association. Vol. 57, No. 1/1991. pp. 69–84.

Smith, Billy G. (1995): Life in Early Philadelphia. Documents From the Revolutionary and Early National Periods. Penn State Press. Harrisburg.

Smith, Neil (1979): Toward a Theory of Gentrification. In: Journal of the American Planning Association. Vol. 45, No. 4/1979. pp. 538–548.

Smith, Neil (1996): The Urban Frontier: Gentrification and the Revenchist City. New York.
Snyder, Martin P. (1975): City of Independence: Views of Philadelphia Before 1800. Praeger Publisher. New York.
Southworth, Susan/Southworth, Michael ([4]1992): American Institute of Architects (AIA) – Guide to Boston. BSA – The Bostonian Society of Architects (ed.). Old Saybrook, Connecticut.
SRL – Vereinigung für Stadt-, Regional- und Landesplanung e.V. (ed.) (1999): New Charta of Athens 1998. Berlin.
Stanton, Phoebe B. ([2]1997): The Gothic Revival and American Church Architecture. An Episode in Taste 1840–1856. John Hopkins University Press. Baltimore.
Starr, Roger (1985): The Rise and Fall of New York City. New York.
Statistisches Bundesamt (ed.) (1995): Länderbericht Vereinigte Staaten 1994. Wiesbaden.
Stevick, Philip (1996): Imaging Philadelphia. Travelers' Views of the City from 1800 to the Present. University of Pennsylvania Press. Philadelphia.
Stipe, Robert E. (1984): Historic Preservation in Foreign Countries. Washington, DC.
Strubelt, Wendelin (1999): Globalisierung – ein neuer Begriff, ein altes Thema? In: BBR (ed.): Informationen zur Raumentwicklung. Heft 1. Bonn. pp. 67–70.
Sullebarger, Beth (ed.) (1985): Historic Preservation: Forging Discipline. Preservation Alumni Inc. New York.
Theodorson, George A. (ed.) ([2]1982): Urban Patterns: Studies in Human Ecology. University Park.
Tiesdell, Steven et al. (1996): Revitalizing Historic Urban Quarters. Oxford, UK.
Timmons, Sharon (1976): Preservation and Conversation: Principles and Practices. Preservation Press. Washington, DC.
Tönnies, Gerd (1990): Tendenzen der Urbanisierung in den Vereinigten Staaten. In: Lafrenz, Jürgen/Nagel, Frank Norbert (eds.): Der nordatlantische Raum. Mitteilungen der Geographischen Gesellschaft in Hamburg. Band 80. Stuttgart.
Toulan, Nohad A. (1970): The Distribution and Value of Open Land in the Philadelphia Area. In: Wallace, David A. (ed.): Metropolitan Open Space and Natural Process. Philadelphia. pp. 53–161.
Trimborn, Jürgen (1997): Denkmale als Inszenierung im öffentlichen Raum. Verlag Ralf Leppin. Köln.
Toll, Jean Barth/Gillam, Mildred S. (eds.) (1995): Invisible Philadelphia: Community Through Voluntary Organizations. Temple University Press. Philadelphia.
Tunnard, Christopher (1975): The United States: Federal Funds for Rescue. In: UNESCO (ed.): The Conservation of Cities. New York. pp. 81–106.
Uhlig, Klaus R. (1977): Vereinigte Staaten von Amerika: Zur zukunftsorientierten Bewahrung erhaltenswerter Städte. In: Stadterhaltung Band 1. München.
UNESCO (ed.) (1996): Die UNESCO-Liste des Welterbes. Bonn.
United States History Atlas (1984). Maplewood, New Jersey.
Urban Archives (ed.) (n. d.): Major Manuscript Collections. Temple University Press. Philadelphia.
U.S. Department of Commerce (ed.) (1972): Census 1970. Bureau of the Census. Washington, DC.
U.S. Department of Commerce (ed.) (1975): Historical Statistics of the United States. Colonial Times to 1970. Part 1. Bureau of the Census (ed.). Washington, DC.
U.S. Department of Commerce (ed.) (1984): Census 1980. Bureau of the Census (ed.). Washington, DC.

U.S. Department of Commerce (ed.) (1985): County Business Patterns 1983. Pennsylvania. Bureau of the Census (ed.). Washington, DC.

U.S. Department of Commerce (ed.) (1993): We the Americans .. Washington, DC.

U.S. Department of Commerce (ed.) (1994): Census 1990. Bureau of the Census (ed.). Washington, DC.

U.S. Department of Commerce (ed.) (1995a): County Business Patterns 1993. Pennsylvania. Bureau of the Census (ed.). Washington, DC.

U.S. Department of Commerce (ed.) (1995b): Statistical Abstract of the United States 1995. Bureau of the Census (ed.). Washington, DC.

U.S. Department of the Interior (ed.) (1980): Census 1980. User Guide. Bureau of the Census (ed.). Washington, DC.

U.S. Department of the Interior (ed.) (1982): Independence. Official Guide to Independence National Historical Park. Washington, DC.

U.S. Department of Interior (ed.) (1987): National Historic Landmarks. Washington, DC.

U.S. Department of the Interior (ed.) (1990a): Congress Hall. Capitol of the United States, 1790–1800. Washington, DC.

U.S. Department of the Interior (ed.) (1990b): The National Parks: Lesser-Known Areas. Washington, DC.

U.S. Department of the Interior (ed.) (1991a): Chesapeake and Ohio Canal. Official National Park Handbook. No. 142. Washington, DC.

U.S. Department of the Interior (ed.) (1991b): How to Apply the National Register Criteria for Evaluation. Instructions. National Park Service (ed.). Washington, DC.

U.S. Department of the Interior (ed.) (1992): Manual for State Historic Preservation Review Boards. Washington, DC.

U.S. Department of the Interior (ed.) (^{2}1994): Standards for Rehabilitation and Guidelines for Rehabilitating Historic Buildings. Washington, DC.

U.S. Department of the Interior (ed.) (1995): Standards for the Treatment of Historic Properties. Guidelines. Pittsburgh.

U.S. Department of the Interior (ed.) (1996): The National Parks: Index 1997–1999. Official Index of the National Park Service. Washington, DC.

U.S. Department of the Interior/National Park Service (eds.) (n. d.): Historic Preservation Certification Application. Instructions. Washington, DC.

U.S. Department of the Interior/National Park Service (eds.) (1990): National Register of Historic Places: Registration Form. Washington, DC.

U.S. Department of the Interior/National Park Service (eds.) (1994): Damaged and Threatened National Historic Landmarks. Report. Washington, DC.

U.S. Department of the Interior/National Park Service (eds.) (1995): The Secretary of the Interior's Standards for the Treatment of Historic Properties with Guidelines for Preserving, Rehabilitating, Restoring & Reconstructing Historic Buildings. Washington, DC.

U.S. Department of the Interior/National Park Service/American Architectural Foundation (eds.) (1987b): The Great American Landmarks Adventure. Washington, DC.

U.S. Department of the Interior/National Park Service/Cultural Resources Programs (eds.) (1993): Federal Historic Preservation Laws. Washington, DC.

U.S. Department of the Interior/National Park Service/Preservation Assistance Division (eds.) (1994): Second Lives: A Survey of Architectural Artifact Collections in the United States. Washington, DC.

U.S. Department of the Interior/National Park Service/Preservation Assistance Division (eds.) (1998): Affordable Housing Through Historic Preservation. Tax Credits and the Secretary of the Interior's Standards for Historic Preservation. Washington, DC.
U.S. Department of Education (ed.) (1996): Urban Schools: The Challenge of Location and Poverty. Washington, DC.
U.S. Department of Labor (ed.) (May 1996): Occupational Compensation Survey. Philadelphia, Pennsylvania, New Jersey Metropolitan Area. Washington, DC.
US-Government Manual (1999): 1998–1999. Washington, DC.
Uzzell, David L. (1989): Heritage Interpretation. Belhaven Press. London.
Venturi, Robert et al. ([15]1997): Learning From Las Vegas. Cambridge, Massachusetts.
Vergara, Camillo José (1995): The New American Ghetto. New Brunswick.
Veyne, Paul (1984): Writing History. Wesleyan University Press. Middletown, Connecticut.
Visitors Guide Map/Charleston (1999): City of Charleston. Charleston.
Wagner, Richard (1991): Local Government and Historic Preservation. In: National Trust for Historic Preservation (ed.): Historic Preservation Forum. No. 2–3/1991. Washington, DC.
Walker, Lester (1997): American Shelter. Woodstock, New York.
Wallace, David A. et al. (eds.) (1970): Metropolitan Open Space and Natural Process. University of Pennsylvania Press. Philadelphia.
Wallace, Mike (1996): Mickey Mouse History and Other Essays on American Memory. Temple University Press. Philadelphia.
Walle, Alf H. (1998): Cultural Tourism: A Strategic Focus. Boulder, Colorado.
Ward, David (ed.) (1979): Geographic Perspectives on America's Past. Readings on the Historical Geography of the United States. Oxford University Press. New York.
Warner, Sam Bass Jr. (1972): The Urban Wilderness: A History of the American City. New York.
Warner, Sam Bass Jr. (1991): The Private City: Philadelphia in Three Periods of Its Growth. Philadelphia.
Weber, Achim (1998): Instrumente und Grenzen des Umgebungsschutzes bei Baudenkmälern. Dissertation Universität Köln.
Webster, Richard ([2]1981): Philadelphia Preserved. Temple University Press. Philadelphia.
Weiler, Conrad (1974): Philadelphia: Neighborhood, Authority and the Urban Crisis. New York.
Weiler, Conrad (1980): The Neighborhood's Role in Optimizing Reinvestment: Philadelphia. In: Laska, Shirley B./Spain, Daphne (eds.): Back to the City. Issues in Neighborhood Renovation. Elmsford, New York. pp. 221–235.
Weinhardt, Carl J. Jr. (1973): The Domestic Architecture of Beacon Hill 1800–1850. The Bostonian Society (ed.). Boston.
West, Cornel (1994): Race Matters. Random House. New York.
Williams, Norman Jr./Kellogg, Edmund H./Gilbert, Frank B. (eds.) (1983): Readings in Historic Preservation: Why? What? How? Rutgers University Press. Piscataway.
Wilson, George (1975): Yesterday's Philadelphia. Miami.
Wilson, William H. (1980): The Ideology, Aesthetics and Politics of the City Beautiful Movement. In: Planning and the Environment in the Modern World. Studies in History, Planning and the Environment. Vol. 1: The Rise of Modern Urban Planning 1800–1914. Oxford, UK.
Wilson, William Julius ([2]1990): The Truly Disadvantaged. The Inner City, the Underclass and Public Policy. Chicago.

Witaker, Graig (1996): Architecture and the American Dream. Three River Press. New York.

Weigley, Russell F. et al. (eds.) (1982): Philadelphia: A 300 Year History. Philadelphia.

Whiffen, Marcus (^{4}1996): American Architecture Since 1780. A Guide to the Styles. MIT Press, Cambridge.

Whiffen, Marcus/Koeper, Frederik (1981): American Architecture 1607–1976. MIT Press, Cambridge.

Wolch, Jennifer/Dear, Michael (eds.) (1989): The Power of Geography: How Territory Shape Social Life. Unwin Hyman Ltd. London, UK.

Wolf, Edwin II (^{2}1990): Philadelphia: Portrait of an American City (a Bicentennial History). Philadelphia.

Wolfram-Seifert, Ursel (1990): Pittsburgh – der wirtschaftliche und städtebauliche Wandel einer amerikanischen Industriestadt zur High Tech – und Dienstleistungsmetropole. In: Lafrenz, Jürgen/Nagel, Frank Norbert (eds.): Der nordatlantische Raum. Mitteilungen der Geographischen Gesellschaft in Hamburg. Band 80. Stuttgart.

Zelinsky, Wilbur (1958): A Method for Measuring Change in the Distribution of Manufacturing Activity: The United States 1939–1947. In: Economic Geography. Vol. 34, No. 2/1958. Worchester, Massachusetts. pp. 95–126.

Zickmund, Susan (1998): Interethnische Konflikte aus amerikanischer Sicht. Rassentrennung und urbane Gruppenbildung in der Migrationsdebatte. In: Heitmeyer, Wilhelm et al. (eds.): Die Krise der Städte. Analysen zu den Folgen desintegrativer Stadtentwicklung für das ethnisch-kulturelle Zusammenleben. Frankfurt am Main. pp. 317–348.

Ziegler, Arthur P. (1971): Historic Preservation in Inner City Areas. A Manual of Practice. The Allegheny Press. Pittsburgh.

Ziegler, Arthur P./Kidney Walter C. (1980): Historic Preservation in Small Towns. A Manual of Practice. Nashville.

Appendix

Glossary

Act
Law that is legally binding. Synonymous with *law*.

Advisory Council on Historic Preservation (ACHP)
Associate advisory council on historic preservation, or federal monument council (see Born 1995, p. 193).

Affordable Housing Program
Term describes "affordable" housing. The particular criteria were established by the HUD:
An apartment is considered affordable if rent expenses for lower and middle-income classes does not exceed 25-30% of income.

Amendment
Constitutional additions to the U.S. Constitution, which was written in 1787.

American Institute of Architects (AIA)
National organization in which architects represent their common interests. The local chapter of the AIA has its own divisions.

Architectural Review Board
A body assigned to the different city planning offices. This board of experts includes architects, historians, city planners, etc., who hold their honorary offices on a voluntary basis. The member board (around 10) decides architectural issues on the preservation of historical monuments and will pass its vote on to the decision-making executive bodies of the City Council. The Architectural Review Board's vote is not binding.
Some of the Architectural Review Boards also wrote the historic district zoning ordinances (or design guidelines) for the cities in question.

Board
A body of experts from the fields of architecture, history, city planning, etc. Depending on the individual state constitution, a board assumes either executive powers or merely advisory functions.

City Council
City Council/Council Assembly with decision-making, executive powers, normally elected by the public for a two-year term. The City Council appoints the mayor, who employs the city manager. As a rule, the city manager hires the city planners.Thus, the City Council tends to be subject to the current political trends of the city's ruling party.

Classical Style
Architectural style with origins in classical Greek architecture, recalling also architecture of Ancient Rome.

Colonial Style
Architectural style which marked and shaped the American colonies during the 17th and 18th centuries. The individual elements borrow mainly from the English style.

Conservation
Conservation is effected to prevent the decomposition of construction material. The term is used in connexion with natural as well as cultural landscapes (Feilden 1997, p. 3): "Conservation must preserve and if possible enhance the message and values of cultural property." (Feilden 1997, p. 6).

Community
A community, also in the sense of the (general) public.

Preservation of Historical Monuments
The process of placing objects, ensembles or districts under historic preservation, in accordance with existing legislation.

Historic Preservation
Procedures involving the restoration and conservation of objects that have already been part of the preservation of historical monuments (in accordance with the legislation). This also includes taking inventory of building material that is eligible for the preservation of historical monuments.

Designation
The designation, labeling and regulation of a historic object as worthy of historic preservation.

Design Guidelines
Established criteria which determine the aesthetic form of urban development. Preliminary guidelines for restorations, architectural styles etc. drafted by the city or neighborhood organization for inhabitants of historic districts.
A design guideline is legally binding if noted or integrated in the historic district ordinance or historic zoning ordinance.

District
A geographically defined area that has a certain character or a particular history (for instance, a large number of historic buildings, a certain ethnic composition).

Division
A (sub-)division, e. g., of an organization.

Fabric (also: Urban Fabric)
Construction material of a building, ensemble, or an area unit.

Federal (Style)
Relating to the Federal level.
Architectural style stemming from the time of the American revolution which resumed with the beginning of the Greek Revival in the early 19th century. It is characterizrd by simple, flat facades with few or no ornamentations.

Federalism
The way in which the federal State is constructed. Contrary to the German usage (*Föderalisten/Förderalismus*), federalists are representatives of a strong central force at federal level.

Flemish Bond
Characteristic stone formation in the stonework with alternations of varnished and matt stones. The shorter and longer stones alternate, to form a decorative pattern on the visible walls.

Government
In American usage, referring to local government as well as the state apparatus (totality of all political institutions).
These multiple meanings gignify that, in the U.S., the state normally represents no abstract-metaphysical entity but rather concretizes personal and material aspects in its exercise of public sovereignty.

Grants-in-aid
Federal subsidies. The federal government provides funds for the identification and nomination of monuments. This kind of financial aid, however, relies on a 50% subsidy from the cities.

Grass Roots Democracy
American interpretation of its power order as a democracy reaching down to the grass roots. This idea reflects the agrarian and romantic tradition of American politics, which involves the notion of self-sufficient units of democracy (family, neighborhood, community).

Gothic Style

Architectural style from the Middle Ages, which first developed in France. This style, mostly utilized in northern Europe, came to the U.S. in the 19th century, where it caused a revival of the ornamental arts and the atmosphere of the Middle Ages.

Greek Revival

An American style that became popular between 1835 and 1850. It brought about a significant break with the Colonial Style and the Federal Style, which had been predominant in America until then. The key elements of this style are taken from the Greek temple and can be found on the front facade.

Hearing

Public hearing. Represents the core element of U.S. parliamentary activity. The hearings serve to inform and discuss the public of legislation or planning activities and act as a forum for the expression of public opinions.

Heritage

Literally: inheritance, inheritance law, genotype.

Regionally specific heritage also refers to cultural, architectural, genealogical and intellectual inheritance.

Historic District

Geographically defined urban or rural area in which a significant concentration of historic buildings or architectural styles exist. The historic district receives special protection through the local historic zoning ordinance, which applies to all buildings, including more recent construction. The individual historic districts in the U.S. vary considerably in size.

Historic District Ordinance (also: Historic Zoning Ordinance)

A local or national law rendering the future conservation of the district possible and establishing in a binding. Owners are therefore legally committed to submit applications to the respective local authorities for any alterations they intend to make to buildings.

Historic Preservation

In U.S. usage, the concept refers not merely to the conservation and preservation of monuments; it also caters for their legal protection.

Historic Rehabilitation Tax Credit

For historic buildings that are registered through the National Park Services, the Tax Reform Act of 1986 allows owners and, in particular cases, leaseholders, a 20% cut in income taxes to compensate for rehabilitation expenses.

Historic Skirt

American attempt to hide the lower floors of new high-rises behind a historic building or ensemble so that the overall historic facade of the street is not disturbed. At the same time, this form is supposed to generate an increase in the density of buildings, particularly in the Central Business District (CBD) of the cities. This method represents a compromise between the proponents of authentic historic facades and the representatives of new development within such areas.

International Style

An architectural style that was developed in the 1920s. This style aims at breaking with some of the cultural and historic elements. This new style intends to cater both to contemporary technology and to the needs of the people. Le Corbusier, Walter Gropius and others were the forerunners of this movement.

Landmark

1. Objects, districts, buildings and structures that are of particular importance. Often, (city, state, etc.) administrative bodies keep a Landscape Register, in which objects are quoted and administered.
2. An object serving as a landmark for spatial orientation in an urban are a (this can be a park or a monument).

Living History
Relatively new method used to respective historic events at their historic places. The visitors play also an important role here.

National Historic Landmark (NHL)
Buildings, structures, historic districts, places and objects of exceptional value, which represent the history and culture of the U.S.A.. The NHL is led by the National Park Service and the U.S. Department of the Interior.

National Register of Historic Places (NR),
Official list in the U.S.A. to list the historic, architectural, archaeological and cultural resources. The NR is led by the National Park Service and the US-Department of the Interior.

Neighborhood Organization
U.S.-specific citizen movement. Titles such as Civic Association or Residents Association are to be understood as being synonyms for a neighborhood organization.

Neo-Classical Style
Architectural style that developed in 18th century Europe. It is based on the rediscovery of the styles of Ancient Greece and Rome.

New Deal
Name for the 1933 government program intorduction by Franklin D. Roosevelt, with whose help the U.S. aimed to overcome the Great Depression from 1933-1945. The concept actually originates from card playing: "new deal" meaning a "new round" of cards.

NIMBY
Acronym for: "not in my backyard".
It refers to protectonists' rejection of undesirableprojects, especially when dealing with socially marginalized individuals or groups.

Postmodern Style
Architectural style originating in the 1960s as a reaction to the sterile and functional architecture of the International Style. Its characteristics often contain exaggeratedly large elements.

Public Welfare
Public interest in the welfare of society. Legitimizes the establishment of the rights of the individual.

Preservation Easements
The imposing of restrictions on owners of houses that have been placed under historic preservation.

Quality-of-life-issues
Subjective sentiments towards elements evident in the immediate environment like noise, vandalism, crime, neighborhood solidarity, etc. Quality-of-life issues are promoted by the neighborhood organizations in particular.

Redevelopment (Authority)
The term describes the development of a newly defined urban area as an answer to the socio-economic reappraisal of the neighborhood. In an effort to coordinate this new development in the U.S., Redevelopment Authorities (RDA) were established in the 1950s.

Reconstruction
1. Exact copy of a building according to historical knowledge.
2. Phase in U.S. history following the Civil War.

Recycling
Drastic adaptation of a building to fulfill a new, contemporary function. Corresponds to the German concept of *Umnutzung*. (Other words: re-use, adaption, etc.)

Renovation
The changing and improving of historic construction materials. Historic elements are reused as much as possible yet channeled towards new purposes. Some elements are exchanged or non-historical materials used.

Restoration
Very elaborate form of historic preservation. Process involves the repair of old, historic material. No adaptations occur to favor modern functions, and no new elements or materials are added.

Segregation
Geographic sorting process of the residential population on the basis of ethnic, socioeconomic or lifestyle-related differences. Upon closer scrutiny, the question of whether segregation is voluntary or forced is crucial.

Urban View
A segment of a city whose aesthetic perception plays a dominant role (for example: skyline).

Urban Landscape
Extensive, comprehensive and abstract term with no evaluative quality per se. Urban landscape involves topography, environments, structures and individual elements of the city (see also: Buchholz 1997, p. 20).

Urban Furnishing
Also: *Enviroelemente*, (Uhlig 1977, p. 10). Railings, bill boards, floor coverings, tree borders, etc. are all urban furnishings. These design elements are meant to contribute to a "cozy" atmosphere.

Street Furnishing
Equipping of public spaces, for example: benches, lights, signposts, emergency-call stations. Street furnishing is maintained with public funds.

Streetscape
Description for streets and their different characteristics.

Urban Renewal
Renewal of the urban area
The urban renewal phase that led to the revitalizing of urban centers in the 1950s and 1960s (Urban Renewal Programs).

U.S. Department of the Interior
Federal department of interior affairs.

Victorian Style
Long, eclectic period marking the reign of Queen Victoria of Great Britain and the social outcomes of her reign (1837-1901).

Visual Pollution
Optical pollution.
Term is often used in relation to one's subjective, rather negative sensations towards one‘s aesthetic perception space.

Vernacular Buildings
Buildings that were not errected under supervision of a trained architect or designer. Also: buildings whose construction is founded on ethnic, cultural or social traditions and less on architectural philosophies.

Windshield-Survey
Recording or surveying done from inside the car. This method is frequently used in areas which show a high crime rate yet have to be recorded statistically.

Zoning Code
Legislation imposed through the local government which decides matters relating to land utilization, building utilization and further aspects of regional development.

Zoning Ordinance
Development plan. Details, standards and legal success vary from city to city.

Appendix A: Tables and Figures A.1 to A.7

Appendix A.1 List of expert interviews

Number	City	Organization	Date	Name	Function/Profession
Iv-Phl-1	Phila-delphia	Mother Bethel Church (Society Hill)	10/7/ 1998	Mrs. Dockins	Public relations
Iv-Phl-2	Phila-delphia	St. Peter Meth-odists Church	10/9/ 1998	Mrs. Lory Floyd	Parish council, lived in Society Hill until 1964
Iv-Phl-3	Phila-delphia	American Indian of Delaware Valley	10/9/ 1998	Mrs. Michele Leonard	Director
Iv-Phl-4	Phila-delphia	--	10/26/ 1998 (T)	Edmund Bacon	Former Chair of the Phila-delphia City Planning Commission
Iv-Phl-5	Phila-delphia	Mother Bethel Church (Society Hill)	10/26/ 1998	Arthur Brownsen	Church member of the AMI
Iv-Phl-6	Phila-delphia	Partners for Sa-cred Places	11/9/ 1998	Tuomi Forrest	Program coordinator
Iv-Phl-7	Phila-delphia	AIA-Philadelphia	11/26/ 1998	Alvin Holm	AIA-Architect
Iv-Phl-8	Phila-delphia	Washington Square West Civic Associa-tion	4/30 1999	John Jorden	Treasurer of neighbor-hood-organization
Iv-Phl-9	Phila-delphia	AIA-Philadelphia	5/7/ 1999	Robert Hotes	Chair of the AIA-Philadelphia Historic Preservation Com-mittee and Associate-Architect at Susan Maxmann & Partner
Iv-Phl-10	Phila-delphia	National Park Service (NPS), Philadelphia	5/10/ 1999 (T)	Cliff Tobias	Park History Program Planner at the National Park Service

Number	City	Organization	Date	Name	Function/Profession
Iv-Phl-11	Phila-delphia	Philadelphia Historical Commission (PHC)	5/18/ 1999	Laura M. Spina	Historic Preservation Planner at the Phila-delphia Historical Com-mission
Iv-Phl-12	Phila-delphia	Philadelphia City Planning Commission	5/24/ 1999 (T)	Michael Roepel	City Planner at the City of Philadelphia
Iv-Phl-13	Phila-delphia	Old City Civic Association (OCCA)	5/24/ 1999	Richard Thom	Chair of Development Committee of the OCCA
Iv-Phl-14	Phila-delphia	Society Hill Civic Association (SHCA)	5/25/ 1999 (T)	Bernice Hamel	Co-chair of the Historic Preservation Division of SHCA
Iv-Phl-15	Phila-delphia	Philadelphia Preservation Alliance for Greater Phila-delphia	5/26/ 1999 (T)	Aliki Strakes	Lawyer of the Preserva-tion Alliance
Iv-Phl-16	Phila-delphia	Philadelphia Historical Commission	5/27/ 1999	Richard Tayler, Ph.D.	City Historic Preservation Officer of the City of Phila-delphia
Iv-Har-1	Harris-burg, PA	State Historic Preservation Of-fice	5/14/ 1999 (T)	Michael Leverfe	Program director of His-toric Preservation Plan-ning
Iv-Bos-1	Boston	Boston Land-marks Commis-sion (BLC)	5/3/ 1999	Ellen Lip-sey	City Historic Preservation Officer, Executive Director of Boston Landmark Com-mission (BLC)
Iv-Bos-2	Boston	Boston Historic Inc.	5/2/ 1999	Wayne E. Beitler	Assistant to the Director
Iv-Bos-3	Boston	Boston Preser-vation Alliance	5/3/ 1999	Albert Rex	Executive Director
Iv-Bos-4	Boston	South Boston Mainstreets, Inc. (Boston Main Streets-Program)	5/3/ 1999	Rosemary J. Powers	Director
Iv-Bos-5	Boston	Gasman (Beacon Hill)	5/4/ 1999	Richard Connolly	Owner of Gas company within Beacon Hill

Number	City	Organization	Date	Name	Function/Profession
Iv-Cha-1	Charleston	Post and Courier (Newspaper in Charleston)	5/17/1999 (T)	Robert Behre	Journalist (specializing in historic preservation and city planning)
Iv-Cha-2	Charleston	Radcliffborough	5/18/1999 (T) and 5/21/1999	Robert Ballert	President of neighbor-hood-Organization
Iv-Cha-3	Charleston	Preservation Society of Charleston	5/20/1999	Robert Gurley	Assistant Director
Iv-Cha-4	Charleston	City of Charleston	5/20/1999	Charles Edwin Chase, AIA	Architect and Preservation Officer of the City of Charleston
Iv-Cha-5	Charleston	City of Charleston	5/20/1999	Brian Sheehan	Ombudsman of the City of Charleston
Iv-Sav-1	Savannah	Historic Savannah Foundation	5/21/1999	Lesley Tucker	Program coordinator
Iv-Sav-2	Savannah	Savannah Historic District Board of Review	3/27/2000 (T)	Mary Reiter	City Preservation Officer of the City of Savannah
Iv-Wash-1	Washington, DC	International Council of Monuments and Sites (ICOMOS)	3/28/2000 (T)	Gustavo F. Araoz	Director of US-ICOMOS

(T)= Telephone interview

Appendix A.2 Interview guide

The following contains the Interview Guide, which was applied individually to the in-depth interviews with experts and organizations, each one in a slightly revized form.

1. Organizational Structure, Budget, Financial Situation

- Honorary activities, full-time employees, (temporary employees)
- Membership modalities (categories, individual members, business members, institutions)
- Membership structure (age, social status, ethnic category, particular interest group)
- Financial resources (membership fees, donations, public funds, endowments or public-money applications)
- Annual budget, budgetary allocations to historic preservation
- Office availability
- Meetings, conventions, public relations, degree of public relations
- Formal functions (hierarchical structure, particular interest groups, interest in historic preservation)
- Areas of legal influence
- Available archives/libraries

2. Objectives and Methods of Activities

- History of the organization
- Main objectives, significance of historic preservation
  - Historic preservation – for what purpose?
  - Economic/cultural interests?
  - Are re-uses/conversion possible in this neighborhood?
- Consulting, service function
- Cooperative/confrontational approach
- Activities that have succeeded
- Identification with local complex (neighborhood/city/SMSA)

3. Self-Perception of Organizations, Cooperative Process

- Evaluation of activities (subjective insights)
- Internal view/external view of organization
- Which problems exist with respect to historic preservation in this city?
- Prospects and future planning
- Cooperation with other groups involved in historic preservation, assessing work by other groups

Karolin Frank♦1998

Appendix A.3 Architectual periods

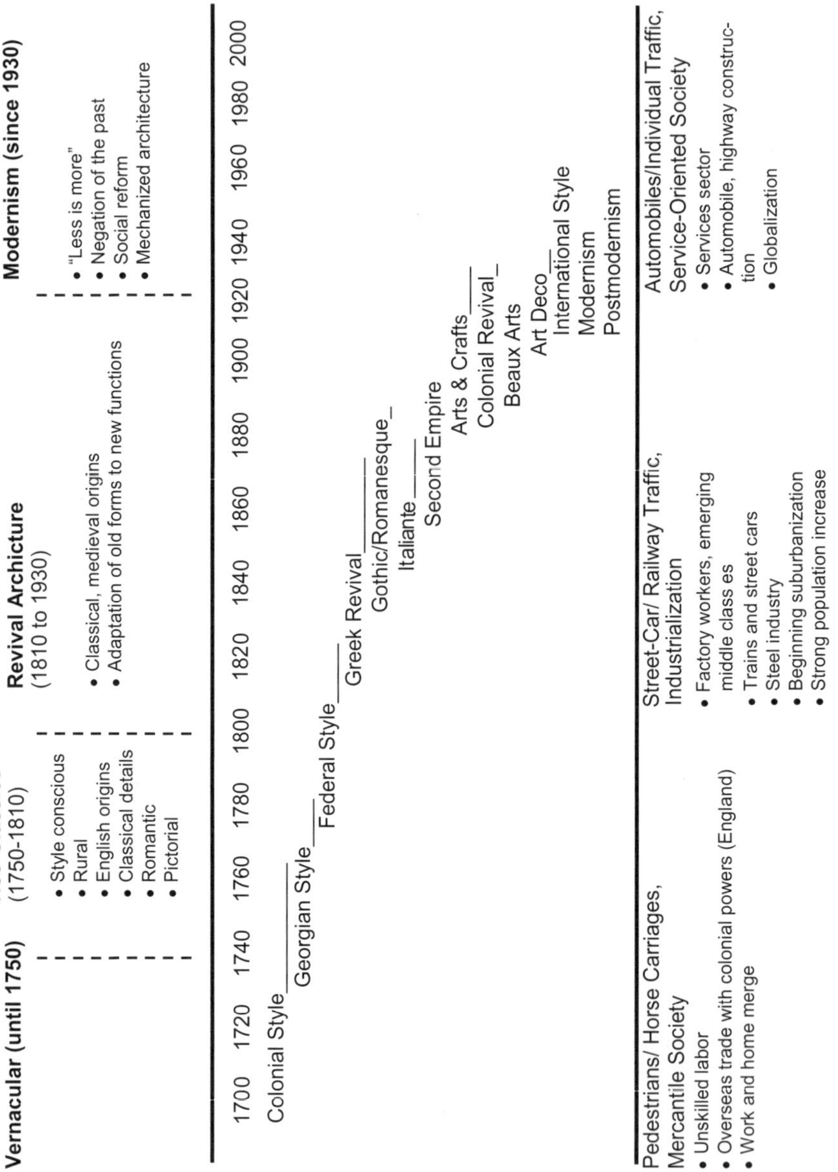

Source: Based on Foundation for Architecture (1991) and Walker (1997)

Appendix A.4 Structuring of requirement groups

In connection with the municipal functions, Hofmeister has identified specific groups based on their affiliations. These groups can be organized by requirement groups (also: *Konsumgruppen*, *Branchengruppen* (consumer groups, affiliation groups), Hofmeister [7]1999, pp. 66). With respect to the subject of this thesis, Hofmeister's structuring of requirement groups can be modified on the basis of the urban structures in the U.S. and the way it is reflected in the historic districts.

1. Public departments, educational institutions, health and social services
2. Official institutions, interest groups, neighborhood organizations
3. Religious communities and their facilities
4. Cultural institutions (theaters, museums)
5. Restaurants, bars
6. Company quarters
7. Banks und insurance companies
8. High-end private, and, or occasions, personal service providers (physicians, attorneys, real estate agents, engineers, tax consultants, advertizing agencies)
9. Basic personal services (hairdressers, dry cleaners, shoe repairmen, plumbers)
10. Retail sector (different requirement degrees)
11. Source: based on Hofmeister (1999, pp. 66)

Appendix A.5 Possible impediments to historic preservation

- Conflicts of interest/competition among participants
- Low level/lack of readiness for conflict
- Varying starting points
- Unintended self-dynamic effects within administration
- Not enough lobbying
- Diverging areas of interest
- Lack of communication among participants
- Aims diverge with respect to a specific neighborhood

Appendix A.6 Registration form

NPS Form 10-900
(Oct. 1990)

OMB No. 10024-0018

United States Department of the Interior
National Park Service

National Register of Historic Places Registration Form

This form is for use in nominating or requesting determinations for individual properties and districts. See instructions in *How to Complete the National Register of Historic Places Registration Form* (National Register Bulletin 16A). Complete each item by marking "x" in the appropriate box or by entering the information requested. If an item does not apply to the property being documented, enter "N/A" for "not applicable." For functions, architectural classification, materials, and areas of significance, enter only categories and subcategories from the instructions. Place additional entries and narrative items on continuation sheets (NPS Form 10-900a). Use a typewriter, word processor, or computer, to complete all items.

1. Name of Property

historic name ______________________________

other names/site number ______________________________

2. Location

street & number ______________________________ ☐ not for publication

city or town ______________________________ ☐ vicinity

state __________ code ____ county __________ code ____ zip code ______

3. State/Federal Agency Certification

As the designated authority under the National Historic Preservation Act, as amended, I hereby certify that this ☐ nomination ☐ request for determination of eligibility meets the documentation standards for registering properties in the National Register of Historic Places and meets the procedural and professional requirements set forth in 36 CFR Part 60. In my opinion, the property ☐ meets ☐ does not meet the National Register criteria. I recommend that this property be considered significant ☐ nationally ☐ statewide ☐ locally. (☐ See continuation sheet for additional comments.)

____
Signature of certifying official/Title Date

____
State of Federal agency and bureau

In my opinion, the property ☐ meets ☐ does not meet the National Register criteria. (☐ See continuation sheet for additional comments.)

____
Signature of certifying official/Title Date

____
State or Federal agency and bureau

4. National Park Service Certification

I hereby certify that the property is: Signature of the Keeper Date of Action

☐ entered in the National Register.
☐ See continuation sheet. ______________________________

☐ determined eligible for the National Register
☐ See continuation sheet. ______________________________

☐ determined not eligible for the National Register. ______________________________

☐ removed from the National Register. ______________________________

☐ other, (explain:) __________

__________ ______________________________

____

Name of Property ______________________ County and State ______________________

5. Classification

Ownership of Property
(Check as many boxes as apply)

- ☐ private
- ☐ public-local
- ☐ public-State
- ☐ public-Federal

Category of Property
(Check only one box)

- ☐ building(s)
- ☐ district
- ☐ site
- ☐ structure
- ☐ object

Number of Resources within Property
(Do not include previously listed resources in the count.)

Contributing	Noncontributing	
		buildings
		sites
		structures
		objects
		Total

Name of related multiple property listing
(Enter "N/A" if property is not part of a multiple property listing.)

____

Number of contributing resources previously listed in the National Register

____

6. Function or Use

Historic Functions
(Enter categories from instructions)

____

Current Functions
(Enter categories from instructions)

____

7. Description

Architectural Classification
(Enter categories from instructions)

____

Materials
(Enter categories from instructions)

foundation ______________________

walls ______________________

roof ______________________

other ______________________

Narrative Description
(Describe the historic and current condition of the property on one or more continuation sheets.)

Name of Property ______ County and State ______

8. Statement of Significance

Applicable National Register Criteria
(Mark "x" in one or more boxes for the criteria qualifying the property for National Register listing.)

☐ **A** Property is associated with events that have made a significant contribution to the broad patterns of our history.

☐ **B** Property is associated with the lives of persons significant in our past.

☐ **C** Property embodies the distinctive characteristics of a type, period, or method of construction or represents the work of a master, or possesses high artistic values, or represents a significant and distinguishable entity whose components lack individual distinction.

☐ **D** Property has yielded, or is likely to yield, information important in prehistory or history.

Criteria Considerations
(Mark "x" in all the boxes that apply.)

Property is:

☐ **A** owned by a religious institution or used for religious purposes.

☐ **B** removed from its original location.

☐ **C** a birthplace or grave.

☐ **D** a cemetery.

☐ **E** a reconstructed building, object, or structure.

☐ **F** a commemorative property.

☐ **G** less than 50 years of age or achieved significance within the past 50 years.

Areas of Significance
(Enter categories from instructions)

______

Period of Significance

______

Significant Dates

______

Significant Person
(Complete if Criterion B is marked above)

______

Cultural Affiliation

______

Architect/Builder

______

Narrative Statement of Significance
(Explain the significance of the property on one or more continuation sheets.)

9. Major Bibliographical References

Bibilography
(Cite the books, articles, and other sources used in preparing this form on one or more continuation sheets.)

Previous documentation on file (NPS):

☐ preliminary determination of individual listing (36 CFR 67) has been requested
☐ previously listed in the National Register
☐ previously determined eligible by the National Register
☐ designated a National Historic Landmark
☐ recorded by Historic American Buildings Survey # ______
☐ recorded by Historic American Engineering Record # ______

Primary location of additional data:

☐ State Historic Preservation Office
☐ Other State agency
☐ Federal agency
☐ Local government
☐ University
☐ Other

Name of repository: ______

Name of Property ______________ County and State ______________

10. Geographical Data

Acreage of Property ______________

UTM References
(Place additional UTM references on a continuation sheet.)

	Zone	Easting	Northing		Zone	Easting	Northing
1				3			
2				4			

☐ See continuation sheet

Verbal Boundary Description
(Describe the boundaries of the property on a continuation sheet.)

Boundary Justification
(Explain why the boundaries were selected on a continuation sheet.)

11. Form Prepared By

name/title ______________

organization ______________ date ______________

street & number ______________ telephone ______________

city or town ______________ state ______ zip code ______

Additional Documentation
Submit the following items with the completed form:

Continuation Sheets

Maps

A **USGS map** (7.5 or 15 minute series) indicating the property's location.

A **Sketch map** for historic districts and properties having large acreage or numerous resources.

Photographs

Representative **black and white photographs** of the property.

Additional items
(Check with the SHPO or FPO for any additional items)

Property Owner
(Complete this item at the request of SHPO or FPO.)

name ______________

street & number ______________ telephone ______________

city or town ______________ state ______ zip code ______

Paperwork Reduction Act Statement: This information is being collected for applications to the National Register of Historic Places to nominate properties for listing or determine eligibility for listing, to list properties, and to amend existing listings. Response to this request is required to obtain a benefit in accordance with the National Historic Preservation Act, as amended (16 U.S.C. 470 *et seq.*).

Estimated Burden Statement: Public reporting burden for this form is estimated to average 18.1 hours per response including time for reviewing instructions, gathering and maintaining data, and completing and reviewing the form. Direct comments regarding this burden estimate or any aspect of this form to the Chief, Administrative Services Division, National Park Service, P.O. Box 37127, Washington, DC 20013-7127; and the Office of Management and Budget, Paperwork Reductions Projects (1024-0018), Washington, DC 20503.

Source: National Trust, October 1990

Appendix B: Tables B.1 to B.9

Appendix B.1 Comparison of the cities' structures: population densities

City	Population[a]	Area (1 mile^2 = 2.5899 km^2)	Population development in the historic district[b]	Population development in the city[c]	Population development in the state[d]
Boston	574,238	48.4 miles^2 = 125.4 km^2	Stable	Slight decrease	+ 5.4% Massachusetts
Phila delphia	1,478,000[e]	129.0 miles^2 = 334.1 km^2	Stable, slight increase	Steady decrease, except in Center City	+ 1.8% Pennsylvania
Charleston	100,122	88.18 miles^2 = 228.3 km^2	Stable	From being stable to a slight decrease	+ 39.1% South Carolina
Savannah	131,647[f]	64.02 miles^2 = 165.8 km^2	Slight decrease	Slight decrease	+ 47.1% Georgia

[a] Sources: In each case latest projections by the city (1998)
[b] U.S. Census 1970, 1980 and 1990
[c] U.S. Census 1970, 1980 and 1990
[d] Statistisches Bundesamt 1995, pp. 41
[e] Fischer Weltalmanach 1999, p. 835
[f] 1998 Census Population estimate for the City of Savannah.
It is striking that the two cities in the sunbelt have experienced a very marked population increase since 1970.

KF ♦ 2000

Appendix B.2 Main characteristics of the historic districts

City	Name of the historic district	Administrative Form	Size of the historic district	Historic function of the district	Present Function/ Characteristic
Philadelphia	Society Hill and Pennsylvania Hospital of Washington Square West	National Historic District (1955) Local Historic District (1999)	0.58 $miles^2$ = 1.5 km^2 (estimated)	Colonial era, residence of the Free Traders	"Urban, residential"
Boston	Beacon Hill Historic District	Local Historic District (1955) through Beacon Hill Civic Association, extended to present borderline: National Historic Landmark (1963)	1.00 $miles^2$ = 2.59 km^2		"Urban, residential"
Charleston	Old and Historic Charleston National Register District (1966)	Established 1931 through Special Commission on Zoning (City of Charleston) 1939: extended by the city; 1966: Extension (tripling of area) National Historic Landmark	7.61 $miles^2$ = 19.7 km^2 (total area of CBD, with varying degrees to which historic building material is preserved and new development is regulated)	Old area of city center	Entire CBD: Living, work, riverfront
Savannah	Savannah Historic District (1966)	National Historic Landmark (1966) National Register (1973)	Savannah Historic District: 0.76 $miles^2$ = 1.97 km^2 Victorian District: 0.26 $miles^2$ = 0.67 km^2	Planned prominent and unique city structure	Entire CBD: Living, work, riverfront

KF ♦ 2000

Appendix B.3 Legislation for the historic districts

City	Name of historic district	Legal basis/ Legislation in place	Design guideline in place
Philadelphia	Society Hill and Pennsylvania Hospital of Washington Square West	No, initiated by Philadelphia Historical Commission (PHC), legislation by PHC, drawn up since establishment of Local Historic District in 1999.	Yes, drawn up by Philadelphia Historical Commission
Boston	Beacon Hill Historic District	Yes, drawn up specifically for Beacon Hill by Initiative of Beacon Hill Civic Association (BHCA)1975	Yes, drawn up by Boston Landmarks Commission in close collaboration with the neighborhood organization, Beacon Hill Civic Association
Charleston	Old and Historic Charleston District	No, but a sound, well-trained Architectural Board, "Charleston Code"	No, is largely covered by the legislation
Savannah	Historic Savannah District	Yes, submitted by the city (1990), since 1964, only 16 guidelines exist for the construction of new buildings	Yes, also prescribing a height index for new development both in the historic district and in adjacent areas

KF♦2000

Appendix B.4 Comparative overview of the conditions for changes

City/ Name of historic district	New construction	Additions	Demolition[d]	Alteration
Philadelphia[b] Society Hill and Pennsylvania Hospital of Washington Square West	Permit required, color, material, height must match with the existing construction	Permit required for all buildings in the historic district, special conditions only for everything that is visible to the public from the street	Permit required for all buildings in the historic district	Permit required for all buildings in the historic district, guidelines apply for any structures that are visible from the street
Boston[c] Beacon Hill Historic District	Permit required, there are precise guidelines for color choice, material, sizes of window frames etc.	Permit required for all buildings in the historic district, strict guidelines apply for any structures that are visible from the street	Permit required for all buildings in the historic district	Permit required for all buildings in the historic district, strict guidelines apply for any structures that are visible from the street

[a] Valid for all cities: Only changes to the exterior will be reviewed by the Board of Architectural Review (BAR). Changes in interior architecture and design will be subject to the historic district ordinances or design guidelines of the cities in specific, exceptional cases only.

[b] Source: PHC - Philadelphia Historical Commission (1997)

[c] Sources: BLC - Boston Landmark Commission (1975) and City of Boston (1997).

[d] It is notable that in all four cities, the official conditions for owners, concerning demolition, are very detailed and precise. This is apparently aimed at finding an eventual buyer or redeveloping the building to be demolished. In fact, several demolitions were prevented in the past by reactive initiatives.

KF♦2000

Continued Appendix B.4

City/ Name of historic district	New construction	Additions	Demolition[e]	Alterations
Charleston[f] Old and Historic Charleston District	Permit is required, guidelines are valid for all structures that are visible from the street, no design guidelines drawn up, aspects of zoning (zoning ordinance, height index) very important	Permit is required for all buildings in the historic district, guidelines are valid for all structures that are visible from the street, the replacement of construction material is particulary subject to strict guidelines, aspects of zoning (development ordinance, height index) very important	Permit is required for all buildings in the historic district, permit will be given out only if a building economicially has been proven "useless", or if leaving the building intact might result in a financial emergency for the owner	Permit is required for all buildings in the historic district, alterations in construction material will be particularly subject to exact control
Savannah[g] Savannah Historic District	Height index has to be respected, material und color as well as form und shape of the new construction should be "in visual alignment with the historic district"	Permit is required for all historic buildings within the historic district, particularly as regards any, change in construction material or color etc. which will be subject to exact control	Permit is required for all historic buildings in the historic district, permit will be given out only if the building has economicially been proven "useless" or if leaving the building intact might result in a financial emergency for the owner	Permit is required for all the buildings in the historic district, the replacement of the construction material, selection of color, etc. will be particularly subject to meticulous control

[e] Sources: City of Charleston and Board of Architectural Review 1996, City of Charleston 1996 a-d.

[f] Sources: City of Savannah 1997 a, b.

[g] For Charleston: All applications and permits are collected in the archive. In their totality, they will then go to form the citie's "desired Charleston character" (City of Charleston 1996d, Article 2, 54-236).

Appendix B.5 The participants in the historic districts: a comparison

City	Activity of neighbor-hood organization	Administrative Form	Potential of historic district in city planning	Projects	Participants
Philadelphia	Strong	Formally controlled by the neighbor-hood	Still an important means, but to a lesser extent	Redevelopment Authority (RDA) "Urban Renewal"	RDA Philadelphia Historical Commission (PHC) Society Hill Civic Association (SHCA)
Boston	Exceptionally strong	Predominantly neighbor-hood-controlled	Weak	Beacon Hill Civic Association (BHCA) Boston Landmark Commission (BLC)	BHCA City of Boston BLC Boston Redevelopment Authority (BRA)
Charleston	Strong in points	Formal and Institutional	Strong	Special Commission on Zoning	City of Charleston The Preservation Society of Charleston
Savannah	Moderate	Formal, Historic Savannah Foundation (HSF) and the City of Savannah	Very strong	Historic Savannah Foundation (active mainly between 1955 and 1968), revolving funds	Private participants City of Savannah

[a] The sequence in which the cities are listed reflects the extent of participation as well as the chronological order of the participants.

KF ♦ 2000

Appendix B.6 Economic structures compared: the four historic districts

	Function/Use	Sales tax	Selected current programs	Preservation type
Philadelphia	Residential (different uses)	7% (Counties of the SMSA: 6%)	Introduction of Local Historic District, establishment of the design guideline, drawing up and establishing of the legislation	Urban Renewal type
Boston	Residential (different uses)	6% (Counties of the SMSA: 5%)	Revitalizing Charleston Street as an economic bonus for tourism and residents, some Affordable Housing Projects, preservation in general, preserving the status quo	Neighborhood type
Charleston	Residential (different uses), mixed-use area begins towards historic district's outer limits	6% (Counties of the SMSA: 5%)	Affordable Housing Projects, debate on zoning regulations: should additional old factories or empty office spaces be converted to apartments?	City Planning type
Savannah	Mixed use	6% (Counties of the SMSA: 6%)	Affordable Housing Projects, extension of tourism	Revolving Fund type

KF♦2000

Appendix B.7 Selected structural characteristics compared: the four historic districts

City	Degree of utilization of converted historic building continuance	Principal objects of conversion/re-use (selection)	Degree of Gentrification	Degree of attractiveness as a residential area
Philadelphia	Low	Schools, old office buildings	High	High
Boston	Low	Old school buildings, corner shops	High	High
Charleston	High	Old office buildings; downsizing of old Colonial Style apartment buildings	At times high, sometimes absent (due to subsidized-housing objects or housing-projects	High
Savannah	High	Old, smaller factory buildings warehouses, market facilities (today: garage), factories converted into stores and residential utilization	Several streets in the center, decreasing towards edges of historic district, very heterogeneous residential population (age, ethnic and social composition)	Average

KF♦1999

Appendix B.8 Tourism in the four historic districts: a comparison

	Frequency of visitors	**Visitor's places of origin**	**Duration of stay of visitors**	**Opportunities to linger in the historic districts**
Philadelphia	Average, mainly on weekends	Mainly regional, Few international tourists	One-day tourism, one or two days	A few museums
Boston	Average, mainly on weekends	High quota of out-of-state and international tourists	One-day tourism, One or two days	Not many in central district, shopping opportunities are on the edge of historic district (especially on Charleston Street)
Charleston	Very high	Very high quota of out-of-state and international tourists	One-day tourism, several days	Few major attractions for visitors in district
Savannah	High	Few international visitors, predominantly regional tourists	One-day tourism, several days	Abundant opportunities - stores, restaurants etc.

KF♦2000

Appendix B.9 Areas of action and challenges in the four historic districts compared

City	Processes	Impedi-ments	Potentials	Problems
Philadelphia	Regulated by city and neighbor-hood organi-zation	Population decrease, no preferred residential region, few jobs	Extension to a con-vention city, inte-gration of Philadel-phia into a visitors network system, re-duction of taxes to create investment incentives and to in-directly promote restoration plans in-directly	Competition with New York and other metropolitan areas, hardly any recurrent tourism, high tax levels, compared to other cities
Boston	Beacon Hill Civic Asso-ciation regu-lates historic preservation	Zoning, densely populated neighbor-hood	The potential real-ized effectively thus far, possibly one could strive to inte-grate minorities into the neighborhood	Retaining the neighborhood's vitality, no "freez-ing up" of the dis-trict
Charleston	City has in-terest in ex-panding tourism	Residents wish to re-duce tourism	Tourism, inhabitants vitalize the down-town area	Affordable housing in the historic dis-trict, development regulations, stu-dents
Savannah	City has in-terest in ex-panding tourism	Little fund-ing, decreasing population in the city	Further decline in tourism, expansion of university pro-grams (thereby re-taining the younger population in the historic district)	Competition with Charleston (1-2 hour's drive away), Atlanta (shopping), New Orleans (im-age)

KF♦2000

Subject Index

Location: http://www.springer.de/geosci/

You are one **click** *away*
from a **world of geoscience** *information!*

Come and visit **Springer's**

Geosciences Online Library

Books

- Search the Springer website catalogue
- Subscribe to our free alerting service for new books
- Look through the book series profiles

You want to order? **Email to: orders@springer.de**

Journals

- Get abstracts, ToC´s free of charge to everyone
- Use our powerful search engine LINK Search
- Subscribe to our free alerting service LINK *Alert*
- Read full-text articles (available only to subscribers of the paper version of a journal)

You want to subscribe? **Email to: subscriptions@springer.de**

Electronic Media

- Get more information on our software and CD-ROMs

You have a question on an electronic product? **Email to: helpdesk-em@springer.de**

Bookmark now:

http://www.springer.de/geosci/

Springer · Customer Service
Haberstr. 7 · D-69126 Heidelberg, Germany
Tel: +49 6221 345-217/218 · Fax: +49 6221 345-229
d&p · 006910_001x_1c

Springer

Printing (Computer to Film): Saladruck Berlin
Binding: Stürtz AG, Würzburg